AF861900

THE AGE OF PITHECANTHROPUS

# THE AGE
## OF
# PITHECANTHROPUS

BY

Dr. Ir. L. J. C. VAN ES
Corr. Member of the Royal Academy of Science Amsterdam
Member of the Board of Natural Sciences Netherlands Indies
Member of the International Committee of Big Reservoir Dams

WITH 11 MAPS AND 4 PLATES

Springer-Science+Business Media, B.V.
1931

Additional material to this book can be downloaded from http://extras.springer. com

ISBN 978-94-017-5674-7 ISBN 978-94-017-5974-8 (eBook)
DOI 10.1007/978-94-017-5974-8

*I dedicate this book to:*

*Professor Dr. E. Dubois*
*Professor Dr. K. Martin*
*Professor Dr. G. A. F. Molengraaff*

*The Author*

## ANALYSIS OF THE CONTENTS

---

[1]) On several maps the Dutch orthography has been used for topographical names (*oe* is to be read as *u*).

[2]) In the index of various maps and sections the word *molluscs* is erroneously spelt with a *k*.

# PREFACE

The authors interest in the problem of the age of Pithecanthropus was first roused by several discussions held with the late Professor Dr. L. Bolk of Amsterdam, during the authors stay in Holland in 1924.

It was again stimulated, after investigations regarding the occurrence of Vertebrates had started, by a visit of the late Professor Dr. William D. Matthew of Berkeley University (then still of the Museum of Natural History of New York), whom the author had the honour to accompany officially on a trip over Java in 1926

The present publishing of the outcome of these investigations is the fullfillment of a promise given to Professor Dr. Elliot Smith of London, Professor Dr. Davidson Black of Peking, Professor Dr. A. N. Burkitt of Sydney and Professor Dr. Shellshear of Hongkong at an unofficial excursion to Sangiran in Northern Suracarta during the 4th Pacific Science Congress held in Java in 1929.

The author wants to express his special gratitude to Ir. A. C. de Jongh, Chief of the Geological Survey of the Netherlands Indies, firstly, for allowing the author — apart of his official work — to continue his investigations regarding the age of the beds in which Vertebrates occur, secondly, for his permission to have the maps and sections prepared for publishing by the Drawing Bureau of the Geological Survey and, thirdly for his allowing Dr. R. von Koenigswald — to whom hereby many thanks are expressed — to assist the author in determining the collections of Molluscs.

During 1930 and 1931 the author was enthusiastically assisted by the Geologists and Mining Engineers of his staff in collecting field evidence. Special thanks are due to Dr. J. M. W. Nash and Dipl. Ing. H. Herold. The author is certainly not less gratefull to Dr. Dipl. Ing. K. G. Schmidt, who even more closely assisted the author during the field work in the Kendeng Mountains and who carefully attended to the laborious work of registrating the collections of rock specimens.

The author expresses his gratitude to Ir. H. Grondijs, who, in giving much of his spare time to assist the author in his official work, thereby freed him, to give his full attention to the problem under discussion.

The author wants to thank Mr. Swens, Chief of the Drawing Bureau of the Geological Survey, for his personal supervision of the drawing of the maps and sections.

In ultimately preparing the collected evidence for publication, the author received valuable information, concerning several details, from Professor Dr. J. H. F. Umbgrove, for which he hereby expresses his special gratitude.

The author is convinced, that after some time, several problems now left open, will want a new discussion. He expresses the hope, that a revived interest in the problems discussed, will lead to the collecting of new field evidence, enabling to give a still more detailed stratigraphy of the Upper Neogene and Pleistocene beds.

The Hague, October 1931.

# CHAPTER I

## INTRODUCTION

With regard to the evolution of mankind three early hominid forms seem to be of utmost importance, viz. *Pithecanthropus erectus*, *Sinanthropus pekingensis* and *Eoanthropus dawsoni*. Some authors add to these: *Homo heidelbergensis*.

It is no wonder that it took rather a long time before the alleged human character of *Pithecanthropus* and *Eoanthropus* was generally accepted. Where presumed human skulls had been found together with artifacts, there seldom remained any reason for doubt, for the presence of cultural objects showed that the presumed human had spiritual powers superior to those of any known living species of animals, even though the stage of civilization was often rather low and inferior to that of the primitive races now living.

Where those objects fail to occur, however, the human character of the skull is to be solely concluded from its anatomical properties, and in the cases under discussion these properties were found to deviate from those of modern man.

1. *Pithecanthropus erectus.*

In 1894 DUBOIS (1894) first published the now world-famous discovery of *Pithecanthropus erectus* in the vertebrate beds of Trinil in Java.

In 1890 part of a lower jaw had been discovered near Kedungbrubus SW of Mt. Pandan, which was immediately ascribed by DUBOIS to some human form. In 1891 the first find in Trinil consisted of a molar. The next month the skull cap was discovered and this was in the original report described as *Anthropopithecus*. The left femur was excavated at a distance of 15 meters in the next year.

In comparing the skull cap with that of any living and fossil species of anthropoid apes DUBOIS afterwards concluded that the brainvolume of the Trinil skull was of considerably larger proportions than the other forms, hence he changed the original name into *Pithecanthropus*. The anatomical features of the femur showing an upright gait, the designation *erectus* was added.

There has seldom been a subject of more controversy. With regard to the presumed human character of the skull cap of *Pithecanthropus*, doubt arose in consequence of the deviations in the anatomical features, there being no specimen of a fossil human skull with such a small brain case. On the other hand the age of the Trinil beds was much discussed and was described by various authors either as Pleistocene or as Pliocene.

Some decades elapsed before *Pithecanthropus* was generally accepted as a hominid. Public opinion was considerably influenced by DUBOIS' (1924) later publication in which he referred the human character of the lower jaw, the molars and the femur, and the big capacity of the brain case (900 cubic centimeters) in comparison with related pithecoid forms.

The remains consist of the skull cap, a left femur, a fragment of the lower jaw and three molars. From the impression of the brains *Pithecanthropus* is to be placed between *Chimpanse* and *Neanderthalman*. The skull form is much nearer to that of man. The femur proves the habit of an upright gait.

The age of the beds was originally described by DUBOIS (1894) as Lower Quaternary or Upper Pliocene. Afterwards on several grounds DUBOIS (1908) presumed an Upper Pliocene age, a point of view not supported by several other authors.

2. *Eoanthropus dawsoni*.

The fragments of the skull of *Eoanthropus dawsoni* were discovered in 1911 or even earlier by DAWSON in a gravel deposit near Piltdown. They were shown to Dr. (afterwards Sir Arthur) SMITH WOODWARD in 1912.

New excavations at the same spot led to the discovery of a lower jaw, which was considered by SMITH WOODWARD as belonging to the same species as the skull. Several well-known anthropologists however doubted this association, holding forth the strongly simian features of the lower jaw, which they ascribed to a *Chimpanse*.

The age of the gravel deposits in which *Eoanthropus* had been found likewise gave rise to some discussion.

According to BOULE vertebrate remains of two different periods seem to be present. The first fauna contains molars of *Mastodon, Stegodon,* and *Rhinoceros,* and is of Pliocene age. It seems however, that these bones are much attacked by wear and probably occur in a secondary deposit, having been washed down from older beds.

The second fauna shows less wear and is described as Pleistocene, this being the real age of the beds. However there are no characteristic forms indicating the exact stratigraphic level of the Quaternary.

With regard to the artifacts found in the beds, a *Chellean* as well as a *Mousterian* stage of civilization seems possible, the former being considered by DAWSON as the more probable.

The age of the Piltdown beds is considered by several authors as Lower Pleistocene. It is to be mentioned however that the data do not suffice to allow of this age being accepted as fully proved.

3. *Sinanthropus pekingensis.*

The first discovery in the Chou Kou Tien formation near Peking merely consisted of some molars, which were ascribed by DAVIDSON BLACK (1927) to a new hominid genus. In 1928 two lower jaws and some skull fragments and in 1929 a skull cap and several teeth were excavated. In 1930 a second skull was recovered by preparation work from collected material.

The age of the beds was considered by TEILHARD DE CHARDIN and YOUNG (1929) as representing the younger part of the Lower Pleistocene, which has been concluded from the accompanying fauna of vertebrates.

The discovery of *Sinanthropus pekingensis* ends some of the controversies with regard to the lower jaw of *Eoanthropus dawsoni* there being no doubt as to the association of the similar simian lower jaw and the rather hominid skull of *Sinanthropus.*

DAVIDSON BLACK (1931) came to the conclusion that *Sinanthropus* in several features more resembles *Pithecanthropus* than the *Neanderthaloid, Rhodesian, modern hominid* or *anthropoid* groups. On the other hand the archaic specialization of *Pithecanthropus* and *Sinanthropus* shows marked contrasts. *Sinanthropus* could not have been far removed from the type of hominid from which evolved both the extinct *Neanderthaloid* and *Rhodesian* forms and the modern *Homo Sapiens.*

*Sinanthropus* is considered by ELLIOT SMITH (1930) as being of even greather importance than *Eoanthropus* and *Pithecanthropus.* He brought forward that in the cave deposit of Chou Kou Tien there is no doubt with regard to the stratigraphical association of the vertebrate fauna and the human remains, whereas *Eoanthropus* and *Pithecanthropus* occurred in river deposits, which always opens the possibility of different faunas having been washed together.

4. *Homo heidelbergensis.*

In 1907 *Homo* (=*Palaeoanthropus*) *heidelbergensis* was discovered in sandlayers at Mauer near Heidelberg and was described by SCHOETENSACK in 1908.

The discovery consisted of an almost complete and very well preserved lower jaw. It is of powerful dimensions and the chin is entirely absent, giving it a simian character. The teeth are human, the canines are small and the molars show the same features as in modern man.

The terrestrial and fluviatile molluscs occurring in the sands of Mauer point out a more continental climate than the present one.

The vertebrate fauna contains *Elephas antiquus, Rhinoceros etruscus,* a species of *Equus* of an intermediate form between the Pliocene *Equus* and the present living *Equus caballus,* and several other vertebrates of less importance. This fauna defines the age as Quaternary.

The comparison with the lower jaw of *Sinanthropus* and *Eoanthropus,* both of those forms showing a more pronounced simian character, gives reason to accept a younger age for *Homo heidelbergensis.*

In comparing the skull cap of *Sinanthropus* and *Eoanthropus* with *Pithecanthropus,* the latter shows more primitive features, which leads to the conclusion that *Pithecanthropus* is an older form in the evolution of mankind. This is confirmed by the character of the accompanying vertebrate-fauna; *Stegodon* and *Hippopotamus* do not occur in the cave deposits of Chou Kou Tien. As to the possibility of these vertebrates being derived from older deposits, the generally well preserved condition of the vertebrate-remains in the Trinilbeds does not support this supposition.

Notwithstanding the new discoveries made, the author holds the opinion that *Pithecanthropus erectus* is still to be considered as the most primitive and probably earliest hominid form hitherto found.

This does not mean that that no new finds are to be expected. The accompanying fossil vertebrate fauna originating from the Asiatic continent, there is always some chance of discovering the ancestors of *Pithecanthropus erectus* in the vertebrate-bearing formations of India or elsewhere in Southern Asia.

As the value of the hominid forms described, with regard to the problem of evolution of man, depends for a great deal upon their relative age, the stratigraphic position of the beds in which they occurred is of utmost importance.

The author has the impression that a more decisive conclusion with regard to the importance of *Pithecanthropus erectus* would have been possible if the age of the Trinil beds had been better established. In the next chapter the author attempts to give a general view with regard to this side of the problem.

# CHAPTER II

## THE PROBLEM OF THE AGE OF THE TRINIL BEDS

The solution of this problem depends for a great deal upon the possibility of demarcating the Tertiary and Quaternary beds in Java. In trying to do so several authors used a different starting point.

1. Orogenic movements.

VERBEEK and FENNEMA (1896) tried to separate the Tertiary and Quaternary beds by assuming a difference in dip. They surmised the folding to be restricted to the Tertiary beds, proclaiming a horizontal disposition of the Quaternary.

They alleged that the vertebrate beds of Trinil and the Kendeng mountains had been deposited for the greater part by the Solo river. Where the beds seemed to occur at too high a level to bear out this solution, they were described as air sediments of volcanic origin (tuffs).

The Quaternary beds were presumed to cover unconformably an unevenly eroded surface of Neogene marls and limestones.

VERBEEK (1908), while strongly advocating the assumption of a horizontal disposition of the vertebrate beds, published the drawing of a section through the Kendeng mountains between Notopuro and Mount Butak.

The vertebrate beds were represented covering unconformably Miocene layers.

Notwithstanding established dips of 15° in the vertebrate beds near Butak, this statement was considered as of no importance and was described as a more local irregularity.

It will be shown hereafter that this conception is based upon wholly insufficient data.

In the geological descriptions subsequent to publications of

VERBEEK and FENNEMA (1896) mention was often made of an unconformity between the marine Tertiary beds of Sonde and the fresh-water strata of Trinil, but generally the facts brought forward are insufficient to confirm this hypothesis.

According to DOZY (1911) the chief vertebrate horizon in Trinil, showing only a slight dip at the extremities, would in the main have shown a horizontal disposition.

The existance of an unconformity between the Trinil and Sonde beds might easily be deduced from the map of Trinil, published by CARTHAUS (1911). But it is to be emphasized that this map gives an entirely inaccurate idea of the demarcation line between the two formations.

On the other hand the dip of the beds between Notupuro, Kedungbrubus and Terbalan, increasing from 1½ to 15°, induced DUBOIS (1907) to proclaim a Pliocene age of the vertebrate series south of the Kendeng mountains.

ELBERT (1907) held the opinion that the lower Kendeng beds are lying conformably upon the Upper Pliocene marine beds.

An unconformity caused by erosion however, was assumed to separate lower and upper Kendeng beds, the latter being presumed to be of lower Pleistocene age.

HILBER (1921) disputed this assumption in proclaiming the possibility of a Pliocene age of the unconformity.

RUTTEN (1927) demonstrated the existence of movements during the Quaternary South of Semarang. Quaternary tuffs of Ungaran-volcano, while unconformably covering Pliocene marine beds, were found to have undergone an upheaval of 100 to 200 meters.

The author discovered SE of Cheribon a corresponding upheaval of Quaternary beds confirming the statements of RUTTEN.

In the authors opinion it should be borne in mind that recent earthquakes, the epicentra of which happen to occur in Java, are the sequel of orogenic movements which caused the big folding of the Pliocene beds.

Arguments derived from the dip of the beds are but of relative and local value and are insufficient to serve as a proof for the age of the beds; though of course it should always be kept in mind that in Quaternary beds strong dips are not to be expected.

2. Volcanism.

VERBEEK and FENNEMA (1896, p. 943) emphasized the increased volcanic activity in the Quaternary period succeeding the folding of the Tertiary beds; they also quoted some proofs of an earlier start of volcanism.

VOLZ (1907) pointed out the rather slightly eroded character of Mt. Lawu-Kukusan. In his opinion this leads up to a relatively young age of the period in which this now extinct volcano had been still active. The Trinil beds having been found to contain boulders of pyroxene andesite identical to that of Mt. Lawu, he concluded the age of these beds not to be older than Lower Diluvial.

DUBOIS (1908) disputed this assumption. Though Mt. Lawu might have produced some of the volcanic products found in the Trinil beds, the bulk of them would have been derived from Mt. Wilis. But even if the end of the eruptive period of both volcanoes should date from a rather recent time, this need not imply a Quaternary origin of volcanic activity.

RUTTEN (1927, p. 114) likewise pointed out that no proofs have been brought forward for the Quaternary origin of Mt. Lawu. On the other hand the pyroxene andesites occurring in the Trinil beds may just as well originate from an older volcano now invisibly covered by the products of the younger Mt. Lawu.

3. River terraces.

ELBERT (1909) asserted the „primary occurrence" of *Stegodon*, *Elephas* and *Hippopotamus* in Upper Diluvial terrace gravel of the transverse valley of Solo river (North of Ngawi).

RUTTEN (1927) rightly disputed this wholly unfounded assertion.

4. Culture remains.

The occurrence of a fossil hearth in the vertebrate beds of Trinil and of bones showing traces of having been handled by man, as asserted by CARTHAUS (1911) and ELBERT (1908), is not scientifically proved and even by these authors is considered as somewhat doubtful.

5. The anatomical features of *Pithecanthropus erectus*.

Up to the time that DUBOIS (1894) published the discovery of *Pithecanthropus* the acknowledged oldest human remains were those of *Neanderthalman*.

Based upon theories which were then still current, DUBOIS considered *Pithecanthropus* as a necessary link in the evolution of man. In consequence of the more simian character of the skull in comparison to *Neanderthalman*, *Pithecanthropus* was computed to be the older of the two. The traces of *Neanderthalman* being known to date from the latest interglacial period, the age of *Pithecanthropus* was alleged to be necessarily older, viz. Upper Pliocene or Lower Pleistocene.

After a lapse of thirty years DUBOIS (1924) of course revised this point of view in accordance with the newer ideas concerning the descent of man.

*Pithecanthropus* now being considered as representing a sideline in the evolution of man, DUBOIS even went so far as to admit that there is no necessity of adopting a greater age than that of *Neanderthalman*.

Nevertheless the author wishes to point out that this admission is to be regarded as only relating to the local occurrence of *Pithecanthropus* in Java. That is to say, *Pithecanthropus* might have been surviving in this distant spot while other more recent mutations of mankind had already sprung into existance in the continent of Asia and Europe. But with regard to the anatomical features of the various forms of mutation there can be no doubt that *Pithecanthropus* is still to be considered as an earlier collateral branch in the pedigree of man.

6. The process of fossilization.

DUBOIS (1908), in comparing several well-known discoveries, came to the conclusion that in Trinil the process of fossilization had reached a more complete stage than elsewhere. The specific weight of the bones was stated to have attained the rather high value of 2.7, which is 35 % more than that of new bone material.

This seems to have been caused mainly by an increase of the lime content, the organic matter, to which the bones owe their dark colour, having been reduced to merely a few traces.

The advanced process of fossilization in Trinil was considered by DUBOIS as an argument for a Pliocene age.

This assumption is not to be accepted as very convincing. The process of fossilization does not depend upon mere time. So many other factors, the lithological character and the grain of the beds, the climate and the circulation of groundwater are of even greater importance. There is reason to accept a more rapid result for the process of fossilization in Trinil, considering the tropical climate and the porosity of the bone-bearing beds.

7. Marine molluscs.

The beds almost immediateley covered by the vertebrate beds in Trinil are composed of sandy marls bearing marine molluscs.

A considerable quantity of molluscs having been collected at Sonde, the stratigraphic level was designated as Sonde beds.

MARTIN (1908, 1919) deduced the Pliocene age from 150 species of molluscs, of which species 53 % are still living.

DUBOIS (1907, 1908), with regard to the age of the Sonde beds, pointed out the uncertainty of their corresponding with stratigraphically well defined Neogene beds in Europe. He even went so far as to doubt the asserted Pliocene age of the Sonde beds, inferring the possibility of a greater age.

Though all later discoveries seem to confirm the Pliocene age, there was something to be said for DUBOIS' disputing the principles for determining the age of the Sonde beds applied by MARTIN. It should not be lost sight of that the fossil molluscs of Java at that time were not comparable to those of Pliocene beds in Europe. Moreover an evident Pliocene age of the Sonde beds is not sufficient evidence to prove a Quaternary age of the Trinil beds, there being room enough to place the latter in the Upper Pliocene.

8. Fresh-water Molluscs.

The Trinil beds contain only *fresh-water Molluscs*, of which the *gasteropods* have been determined by MARTIN-ICKE (1911). Out of a total of 16 species, two were found to be undeterminable, while 14 all belong to recent species. The percentage of living species thus amounts to 87½, which led to the conclusion of a Quaternary age of the Trinil beds.

HILBER (1921) disputed this assertion and mentioned the figures given by KOBELT in 1894 concerning the extinct species in the Pliocene of Tarente. Out of 260 *marine Molluscs* only 3% were found to be extinct, while all the 21 *fresh-water Molluscs* appeared to belong to still living species.

For this reason he regarded the Quaternary age of the Trinil beds as still unproven. Including two lamellibranchiates, the Trinil beds contain 18 *Molluscs* of which 15 or 83% belong to still living species.

9. Plant remains.

According to SCHUSTER (1911) all the 52 species of plants from the Trinil beds still occurred in Java.

Whereas a Pliocene flora usually consists of *a.* extinct species, *b.* conformable species in remote regions and *c.* species still occurring in the vicinity, the flora of Trinil was to be considered as Diluvial, as neither extinct nor conformable species occur.

HALLIER (1911) maintained that the determinations of SCHUSTER were uncertain and that the comparison of specimens of recent plants was erroneous.

This led HILBER (1921) to state that the Quaternary age of the Trinil flora was not proved.

10. Vertebrates.

MARTIN (1884, 1886, 1888, 1890) was the first to describe vertebrate remains from Java, originating from Mt. Patihajam, Ngembak hill and Surakarta.

As at that time the only known comparable vertebrate fauna was that of Siwalik, the age of these two faunas was held to be the same. This point of view was at first still maintained by MARTIN (1900) with regard to the fauna of the Kendeng or Trinil beds containing *Stegodon* and *Pithecanthropus.*

The determination of the Trinil fresh-water molluscs by MARTIN and ICKE (1911) caused opinions to be changed and afterwards MARTIN (1919) concluded a proved Quaternary age for all occurrences of land vertebrates in Java.

DUBOIS (1892) on the contrary, began to infer a Pleistocene age of the fauna from Trinil, the Kendeng mountains and Patihajam. In 1907, however, he held the Trinil beds to be Pliocene, though somewhat younger than Siwalik, the fauna lacking all older types of Siwalik and of the European Pliocene. In 1908, when discussing the comparison of the Trinil fauna with Siwalik and Narbada, DUBOIS came to the conclusion that the resemblance to the recent fauna is not greater than that to Siwalik and that the conformity is certainly less than between Pleistocene and recent faunas in Europe. This led up to the assumption of a Pliocene age of the Trinil or Kendeng fauna.

FRECH (1904) held the difference between the Trinil and recent fauna in Java to be greater than for Narbada and therefore took Trinil to be the older of the two.

STREMME (1911) described the *Mammals* of the Selenka expedition, excepting the *Proboscidea* which have been determined by JANENSCH (1911).

Notwithstanding a certain conformity with the Pleistocene fauna of Narbada, he proclaimed the probability of a Pliocene age for the Trinil fauna.

In a second publication STREMME (1911) compiled a list of the whole known fauna from Trinil and the Kendeng beds. DUBOIS' collection is mainly derived from the Kendeng mountains, whilst that of the Berlin museum came from Trinil, three species from Kendeng mountains excepted.

According to STREMME's conclusions the fauna differs widely from that of the present time.

None of the forms would be identical with living species, although *Rhinoceros, Buffelus, Bibos, Cervulus* and *Sus* were to be considered as the ancestors of the now living species in Java.

Of the 14 species from the Selenka expedition, 7 species (50%) are extinct in Java and 4 species (28.6%) are wholly extinct.

Of the DUBOIS collection of 19 species, 9 (47.4%) are extinct in Java and 4 (21.5%) wholly extinct.

In comparing the Pleistocene fauna of Asia, Europe, America and Australia, STREMME found less extinct species in the Pleistocene of Europe and India, but the reverse for America and Australia. Altogether the fauna of Trinil corresponds more or less with the Lower Pleistocene fauna of Narbada. But in Narbada out of 14 species only 5 (35.6%) are extinct in India and 3 (21.4%) wholly extinct.

HILBER (1921) doubted the Pleistocene age of the Narbada fauna, disputing the value of the discovery of an early palaeolitic axe described by MEDLICOTT and BANFORD. The Narbada beds were computed to be Upper Pliocene and of the same age as the fauna of Montpellier.

The Trinil beds showing no forms identical with Siwalik, he conceived them to be younger than Narbada, but still of Pliocene age.

DIETRICH (1924) discussed the *Elephantoids*, the evolution of which is well-known and which contains many forms. It seemed to him that of the Javanese forms only *Stegodon airawana-trigonocephalus* were servicable.

In studying the last molars of the lower and upper jaw he deduced the following order of succession.

*Stegodon Clifti* (+ *bombifrons*)
↓
*Stegodon insignis* (+ *ganesha*).
↓
*Stegodon airawana* (+ *trigonocephalus*)

*Clifti* and *bombifrons* occur in the Dhok-Patan and Tatrot zone, against *insignis* and *ganesha* in the Pinjor and upper conglomerate zone of the Siwalik, the age of Dhok-Pathan and Tatrot being Middle Neogene, that of Pinjor Upper Pliocene (Villa-franca) and the upper conglomerates of the upper Siwalik Pleistocene.

*Stegodon orientalis* described by MATTHEW and GRANGER from the Sz'tschwan (China) much resembles the Javanese *Stegodon airawana*, but the latter shows signs of a higher specialization and should be younger.

The farther *Stegodon* scattered from its original region of extension, the higher it was bound to become specialized. As only highly specialized types are recorded from Java, the Philippines and Japan, it would be justified to deduce that the herds which strayed farthest to the NE., E. and SE. are geologically the youngest species of *Stegodon*.

According to Dietrich the separation of the East-Asiatic and East-Indian islands from the Asiatic continent occurred during the Pleistocene. In asserting *Stegodon airawana* to show traces of insular nanism he concluded that the Trinil beds are of quite young Pleistocene age.

It should be kept in mind, however, that in this theory much depends both upon the period in which the islands were separated from the mainland and upon the question of the above mentioned traces of insular nanism. These assertions are not sufficiently proved yet to be generally accepted.

11. Climate.

From the plant remains found in Trinil ELBERT (1908, 1911) deduced a former cooler climate. He stated that these plants occur at the present time in a much higher region and even alleged the temperature to have been 6—8° Celsius lower for the period of the lowest beds and 3—6° Celsius for the vertebrate bed, assuming the plants from the upper layers to occur at the present day at a height of 1500—2000 M. above sealevel.

SCHUSTER (1911) similarly held the opinion that the Trinil flora as a whole would correspond to the present flora from a height of 600—1200 M. above sealevel and thus deduced a lower temperature for the climate during the formation of the Trinil beds. Further he presumed the climate to have been much more pluvial.

Considering that the plant remains have been washed down by rivers from the slopes of the volcanoes, the author is of opinion that the assumption of a cooler climate during the formation of the Trinil beds is unfounded.

CARTHAUS (1911), without mentioning any facts, likewise assumed a wetter climate.

BLANCKENHORN (1911) in a summary assumed the Trinil beds to have been formed during a pluvial period, corresponding to one of the European-American ice periods.

RUTTEN (1927) in summing op the various unsolved problems concerning the age of the Trinil beds, pointed out the insufficient knowledge of the geology of the beds, without which it seems to be impossible to come to an indisputable conclusion.

Considering the cooperation of so many eminent scientists with regard to the determination of fossils from the Trinil beds, it is a rather humiliating confession that the geology of the vertebrate beds in Java is insufficiently known.

The author, in publishing the result of his investigations, aims at filling the deficiencies of our knowledge in this respect and endeavours to confine the solution of the problem of the age of the Trinil beds within narrower limits.

# CHAPTER III

## THE DISTRIBUTION OF FOSSIL VERTEBRATES IN JAVA

1. Bodjonglopang. (Res. W. Priangan).

In March 1929, the author was shown (by Mr. A. W. R. KERKHOVEN) a fossil molar of *Elephas*, excavated in a cave in Miocene limestone near the rubber estate of Panumbangan in the vicinity of Bodjonglopang. This find is of great interest, as a cave deposit generally gives more reliable data in comparison with river deposits. Moreover up to now this is the most Western discovery made in Java. A further investigation may probably lead to remarkable results.

2. Tjitarum. (Res. W. Priangan).

STEHN and UMBGROVE (1929) mentioned the occurrence of fossil vertebrates near Banuradja on the banks of the Tarum river in the central part of the plain of Bandung. They recommanded utmost caution as to conclusions regarding the age of the beds. The dark sandy, argillaceous vertebrate layers, dipping about 30° SW., contain pebbles of loose corals, most probably derived from the subterraneous and neighbouring Neogene limestone. The vertebrate bones might have been transported in the same way.

On the other hand some molars of *Rhinoceros* collected from the same region by Mr. H. GRONDIJS in 1930 show no wear and, though found apart, occur in such a way as to leave no doubt about their having belonged together. This proves that at least some of the vertebrates had been living at the time of the formation of the beds. The bones found by STEHN and UMBGROVE belong to the genera *Cervus*, *Sus* and *Bos*. The absence of *Elephantoids* points to a younger age than the other occurrences of vertebrates.

3. Baribis. (Res. Cheribon).

In 1926 the author discovered part of a *Stegodon* molar in steeply-dipping conglomerates in the vicinity of Baribis, N. of Madjalengka and when revisiting the spot in 1930 he excavated a molar of *Cervus* and some bones. The description of the geology will follow in another chapter.

4. Kromong mountains (Res. Cheribon).

In 1926 a complete *Stegodon* molar was reported to have been found loose on the surface of the low Pliocene hills N. of Kromong mountains. There is no evidence of the stratigraphic horizon from which the molar was derived.

5. Tjidjurai (Res. Cheribon).

In 1926 the author discovered fossil vertebrates in an exposure of Pliocene beds containing molluscs, which had been first discovered by C. A. DE JONG in 1924 and described by MARTIN (1926).

A section through the mollusc-bearing beds compiled by the author was published by 't HOEN (1930).

6. K. Glagah near Bumiaju. (Res. Tegal).

The occurrence of fossil vertebrates, discovered in 1923 was described by VAN DER VLERK (1923). In June 1925, when visiting the spot, the author got the impression that more and better preserved material might be obtainable by excavating the vertebrate beds. However he was not allowed to carry out any further investigations and was unable to lead the proposed excavations which were started by others in November of the same year.

ZWIERZYCKI (1926) found the total thickness of the beds to be 1200 M.

The occurrence of *Mastodon* (*longirostris?*) described by STEHLIN (1926) was regarded by ZWIERZYCKI as a proof for the Pliocene age of the vertebrate beds, which contain fresh-water molluscs and conformably cover mollusc-bearing marine marls.

A section from a survey of ZWIERZYCKI made in 1926 was published by 'T HOEN (1930).

TER HAAR (1929) published a map and a description of the Bumiaju region.

A short description of the vertebrate beds of Bumiaju was published in the Yearly Report of the Java geological survey of 1929 (1930).

7. Madjenang (Res. N. Banjumas).

In 1930 Dr. VAN HOUTEN discovered fossil vertebrates on the southern slopes of the Tertiary mountain range N. of Madjenang.

8. Idju (Res. S. Banjumas).

VERBEEK and FENNEMA (1896, p. 389) mentioned fossil vertebrates occurring at a depth of 4 M. in a loose sandy clay near the Western aperture of the railway tunnel at Idju.

9. Sentolo (Gov. Jogjakarta).

The famous Javanese painter RADEN SALEH (1867) excavated some fossil vertebrate remains near Kalisono and Banjuganti SE. and W. of Sentolo. This collection together with specimens from other places was sent many years afterwards to the Geological Museum of Leiden University. The vertebrate-bearing bed was described by RAHDEN SALEH as a lime-bearing soil covering a soft sandstone.

VAN DIJK (1867) called it a bed of limestone, while VERBEEK and FENNEMA (1896, p. 349) characterized it as a weathered volcanic sand. The latter did not succeed in finding new vertebrate remains, neither did the author get any better results when visiting the spot in 1926.

10. Tjandihill, Semarang (Res. Semarang).

In 1915 Dr. J. HENGEVELD was reported to have excavated some remains of vertebrates in gravels and sandstones forming the northern slopes of Tjandi hill.

The age of the slightly N.-dipping bed is certainly younger than the Pliocene mollusc-bearing marls occurring near Gombel S. of Tjandi, which have been described by MARTIN (1900, 1919).

11. Ngembak hill (Res. Kudus).

MARTIN (1883) described the find of a *Sus* molar, part of a tooth of *Hippopotamus* and a bone of *Elephas*.

The surrounding region being inundated at the time that the author visited the spot in 1930, no geological investigations could be made. It seems, however, that there are signs of an important unconformity between the steeply folded Miocene beds and the slightly-dipping overlying vertebrate series.

12. P u r w o d a d i (Res. Kudus).

VERBEEK and FENNEMA (1896, p. 280) mentioned the discovery of small molars and teeth of vertebrates at a depth of 12 M. below the surface in the centre of the village.

13. P a t i h a j a m. (Res. Kudus and Rembang.)

JUNGHUHN (1857) described Mt. Patihajam as a Tertiary folded mountain, contrary to the opinion of others, who held it to be a small volcano on the southern slopes of the larger volcano Mt. Muriah. He excavated several vertebrate bones, afterwards sent to Leiden.

MARTIN (1884) published the discovery of a molar of *Stegodon*. MARTIN (1888) described two (?) different skulls of *Stegodon*, part of a molar of *Mastodon?*, molars of *Elephas* and *Sus* and remains of *Bison* (?), *Bos* and *Cervus*, all taken from JUNGHUHN's collection.

VERBEEK and FENNEMA (1896, p. 259—261), contrary to the statements of JUNGHUHN, described Mt. Patihajam as a small separate volcano. The vertebrate beds were alleged to be water-deposited tuffs of this volcano.

A stratigraphic section from a survey by the author was already published by 'T HOEN (1930). In another chapter a more complete description will be given.

14. T j e p u (Res. Blora).

The vertebrate series form the subsoil of the small township of Tjepu, as has already been stated by DUBOIS in 1907, and may be continued westward to a distance of more than 5 Km. Near the small campong of Kalen the author discovered in 1930 vertebrates occurring in a conglomerate mainly consisting of pebbles of Miocene limestone and of thickly shelled Pliocene molluscs showing much wear. The bed shows a dip to the south and unconformably covers slightly steeplier dipping Pliocene argillaceous beds. The vertebrate bed in the plain south of Kalen is nowhere thicker than 15 M., as has been proved by drilling.

Farther to the south there occur sandstones of volcanic origin substituting the limestone conglomerate. It seems that the latter was formed by denudation products from the northern Tertiary hills, while the volcanic sandstone was brought down by rivers from a southern origin.

15. Northern border of the Kendeng mountains. (Res. Blora, Bodjonegoro, Modjokerto, Surabaja).

*a*. Randublatung.

In 1927 the author discovered an occurrence South of Randublatung which appeared to be very rich in vertebrates. The beds mainly consist of sandstone and gravel of volcanic origin and overlie the Miocene hills. There is a very pronounced unconformity between the slightly N.-dipping vertebrate beds and the steeply folded Miocene marls.

*b*. Tinggang.

DUBOIS (1907) first mentioned the occurrence of vertebrates. The present author collected a large number of vertebrate remains in 1926—1927 from gravel beds mostly containing pebbles of volcanic origin. Owing to the bad exposures no data were obtainable concerning the relations to the underlying Tertiary beds. In the vertebrate beds in several spots a slight dip to the N. not exceeding 5° was established.

*c*. Dander.

North of the probably Pliocene limestone hills in the vicinity of Dander the author collected in 1927 several loose-lying fossil bones. The only exposures consisted of Pliocene argillaceous beds, but it seems to be quite possible that the bones were derived from presently denudated vertebrate beds.

*d*. Bareng-Tondomuljo

The discovery of vertebrate beds occurring in this region was made by the author in 1927. The region was revisited in 1930 and the geology will be described in another chapter.

Vertebrates were found to occur both in the Pliocene marine beds and in the overlying characteristic volcanic sandstones and gravels, corresponding to the vertebrate beds of Tinggang and Randublatung.

16. Southern border of the Kendeng mountains. (Res. Surakarta, Madiun, Kediri, Modjokerto and Surabaja).

*a*. Sangiran.

Strictly speaking, Sangiran does not belong to the vertebrate series of the southern border of the Kendeng mountains. It is situated more to the south in the plain of Surakarta about

3 Km. E. of Kalioso and 10 Km. N. of the town of Surakarta.

The exposures of the vertebrate beds cover a region of several square Km., of which Sangiran forms the centre.

SCHMUELLING (1864) first mentioned the occurrence of fossil vertebrates near Kalioso.

VERBEEK and FENNEMA (1896, p. 324—326) gave a wholly inadequate description of the geological conditions.

DUBOIS (1907) mentioned Sangiran in the enumeration of occurrences of fossil vertebrates.

The region was thoroughly surveyed by the author in 1927 and will be described in another chapter.

*b*. Baringinan.

In many respects the stratigraphical conditions resemble those of Sangiran, which is not strange; Baringinan is situated only 10 Km. N. of Sangiran and forms the true southern border of the Kendeng mountains. The discovery of fossil vertebrates was made by the author in 1930 and the region was surveyed in that year and the next.

c. Solo river between Gesi and Ngawi.

DUBOIS (1892, 1894) discovered the occurrence of fossil vertebrates at Trinil in 1890 and published a description of the vertebrate beds.

VERBEEK and FENNEMA (1896, p. 249 and 345) dealt with the occurrences of vertebrates at Tawang and Trinil, which they assumed to lie horizontally and unconformably upon the marine tertiary beds.

In the Report of the Selenka expedition CARTHAUS (1911) and DOZY (1911) described the neighbourhood of Trinil.

The author (1929) published a map of the vertebrate series between Sragen and Ngawi, which has now been revised as the Tertiary beds are insufficiently represented.

*d*. The region from Ngawi to Redjuno.

The occurrence of fossil vertebrates near Teguan, Dero and Redjuno was first mentioned by A. BARON SLOET VAN OLDRUITENBERGH (1858) and was investigated by DUBOIS (1907) in 1889—1894.

A survey was made by the author in 1927 and 1931.

*e*. Kedungbrubus—Kedunglembu.

RADEN SALEH (1867) visited the occurrence of fossil vertebrates SW. of Mt. Pandan. DUBOIS (1907) made mention of his researches in this region.

VERBEEK and FENNEMA (1896, p. 249) dealt with this subject in combination with other occurrences in Northern Madiun.

VERBEEK (1908) gave a full description of his more recent investigations with regard to a section through the vertebrate series from Notopuro-Kedungbrubus-G. Butak.

In 1927 and 1930 the author made an extensive survey of the whole region.

*f*. T r i t i k—B a n g l e.

DUBOIS (1907) discovered the occurrence of vertebrates near Tritik and Bangle.

VERBEEK and FENNEMA (1896, p. 170) published a short description of this region.

The author surveyed the principal occurrences in 1927 and 1930—31.

*g*. P u t j a n g a n—S u r u h—K e d a m e a n.

In the vicinity of Mt. Putjangan and Tjoepak remains of fossil vertebrates were collected by the author in 1927, whilst in the same year Dr. J. M. W. NASH discovered the occurrences SE. of Suruh, Kedungpalang and Dadapan.

COSYN (1931) published the discovery of fossil vertebrates in the same region near Gondang (= Kedungpalang) and Djuwet. Probably one of these occurrences corresponds to the one mentioned by RUTTEN (1927) West of Surabaja.

# CHAPTER IV

## PRINCIPLES CONCERNING THE STRATIGRAPHICAL DIVISION OF THE TERTIARY AND QUATERNARY BEDS IN JAVA

1. Foraminifera.

Following the results of previous authors, DOUVILLÉ, RUTTEN etc., VAN DER VLERK (1925) published a stratigraphical division of the tertiary based upon genera of *Foraminifera.*

It was superseded however by one of VAN DER VLERK and UMBGROVE (1927).

Though the authors refrained from indicating the usual divisions of the tertiary, it appears from a later publication by UMBGROVE (1929) that the Eocene corresponds to Tertiary a and b, whilst the Oligocene is represented by c and d and part of the Miocene by e and f.

The use of these genera of *Foraminifera* as guide-fossils appears to be very valuable for the correlation of the Lower Tertiary beds in Europe and the Malay Archipelago.

UMBGROVE (1929) pointed out that the Lower Tertiary of Java possesses genera of *Foraminifera* that are also known from Europe and America, which leads to the supposition that at the time of their first appearance in Europe and in the Indo-Pacific region (the Lower Eocene) *the seas were still connected.*

After a *short separation* during the Upper Eocene, proved by MARTIN (1914) by means of species of *Molluscs,* it seems that a *second connection* existed during the Oligocene, which is concluded by UMBGROVE from the established penetration of European genera of *Foraminifera* into the Indo-Pacific region.

In the Neogene of the Indo-Pacific, new genera of *Foraminifera* originated, and from this fact, together with the separate evolution of genera imported from Europe during the Oligocene, UMB-

GROVE was inclined to suppose that the *sea connection* with Europe was *definitely cut off*.

Later on it will be shown that this *separation was not absolute* during the Lower Miocene. Still, as the evolution of the *Foraminifera* was quite different in the Malay Archipelago and in Europe, these *Foraminifera* cannot serve as a means of identification of contemporaneous Miocene beds.

| Age | | Assilina | Pellatispira | Nummulites | Orthophragmina | Lepidocyclina | Miogypsina | Trillina Howchini | Alveolinella rec. Type | Alveolinella bontangensis | Alveolina | Flosculina | Cycloclypeus | Spiroclypeus | Heterostegina |
|---|---|---|---|---|---|---|---|---|---|---|---|---|---|---|---|
| Recent | | | | | | | | | + | | | | + | | + |
| Quaternary | | | | | | | | | + | | | | + | | + |
| Tertiary | | | | | | | | | + | | | | + | | + |
| | f | | | | | + | + | + | + | + | | | + | | + |
| | e | | | | | + | + | + | | | + | | + | + | + |
| | d | | | + | | + | | | | | + | | ? | | + |
| | c | | | + | | | | | | | + | | | | + |
| | b | | + | + | + | | | | | | + | | | | + |
| | a | + | + | + | + | | | | | | + | + | | | + |

With regard to the uppermost Miocene and Pliocene beds, the *Foraminifera* discussed are of no use at all, the genera and even the species occurring in those divisions appearing to be still living.

On the other hand it seems that the small *Foraminifera*, though for the greater part having a large vertical distribution, still give a possibility of an exact correlation of the Neogene beds by means of a few species which do possess stratigraphical value.

For Java the only study existing up to the present is that of KOCH (1923) with regard to Upper Miocene or Pliocene beds of Kabu, which will be discussed in another chapter. In this respect a tremendous amount of work remains to be done, however, and at present it is rather difficult to come to indisputable conclusions.

A stratigraphic division based upon species of *Foraminifera* was published by GERTH (1929). There being no typical species for the uppermost part of the Miocene and the Pliocene, the identification of those divisions is impossible. Moreover it must be mentioned that in view of the insufficiency of the knowledge of the geology of Java, owing to a complicated structure of the folded Tertiary and to gaps in the sedimentation, great caution is to be excercised in the use of this system.

In the author's opinion the results published are not to be taken as unquestionable.

TAN SIN HOK (1930) in continuing older researches of MARTIN and VAN DER VLERK regarding the structure of the embryonal apparatus of *Cycloclypeus*, proved the evolutionary sequence of species of *Cycloclypeus* to be of interest for a stratigraphic division of the tertiary. He showed the succession for two groups of *Cycloclypeus*, the evolution being characterized by a decrease in the number of the heterostegoid rings.

The *biostratigraphic* division of the tertiary possesses great advantages over that based upon *guide fossils*. The succession of the beds is a fact and gaps are demonstrated by missing links in the chain of descent.

But the application of the system is not so easy; it seems that older generations survived and are found to accompany the younger mutations. In practice it seems to be necessary to count the specimens of all mutations present, and in order to obtain sufficiently reliable results a great number of specimens is required. This premises a special lithological facies of the fossiliferous beds, which restricts the use of the system to special cases.

Moreover a division of the Upper Miocene, Pliocene and Quaternary beds cannot be attained, as this group of layers embraces a single phase of evolution of the only surviving species of *Cycloclypeus*.

2. Marine Molluscs.

MARTIN (1914) pointed to a separate evolution of the marine fauna during the younger Tertiary in the Malay Archipelago as compared to that of Europe, due to a disconnection of the respective seas, which he supposed to have started in the Upper Eocene. It is now known that this separation during the Upper Eocene was only temporary:

UMBGROVE (1929) showed the penetration of Oligocene European genera of *Foraminifera* in the Indo-Malay region.

MARTIN (1931) quite recently showed an *incomplete connection* to have existed between Europe and Java during the Lower Neogene, which, however, was severed again in the Upper Neogene. This statement is based upon the publication by VREDENBURG (1925—1928) concerning the fauna of the post Eocene Tertiary formation of North Western India, in which a „widespread temporary oceanic connection" between Europe and India was alleged to exist during the Middle Oligocene. This connection was supposed to be „perhaps completely interrupted" in the Upper Oligocene, but „to have been reestablished only imperfectly" in the Lower Miocene, whilst a definite separation followed in the Upper Miocene.

MARTIN showed more than 30 Javanese species of *Molluscs* to occur in the Lower Miocenc beds of N.W. India and pointed out the similarity of nearly 40 species of Java and N.W. India in the Upper Miocene and Pliocene beds. As a connection during the Miocene and Pliocene between the seas of Java and N.W. India has been definitely proved, and meanwhile a similar though „imperfect" connection between Europe and N.W. India was supposed to exist during the Lower Miocene, there is no conclusion left but that of MARTIN concerning an *incomplete connection* between Europe and Java during the Lower Miocene.

This conclusion will certainly be of great help in correlating Tertiary beds in Europe and Java.

At the time, however, that MARTIN began to describe the *Mol-*

*luscs* of the Tertiary beds in Java the facts only showed the Tertiary *Mollusc*-fauna of the Indo-Malay region to have followed a separate evolution.

To overcome this difficulty MARTIN proposed to apply similar principles as had already been in use in Europe to obtain a stratigraphic division of the Tertiary beds. By establishing the ratio of still living species of *Molluscs* he attained some sort of division and was able to distinguish certain horizons of the Tertiary.

Excluding some uncertain determinations, the occurrences published on page 27 were established.

As will be shown hereafter in the description of some occurrences where the author succeeded in determining a great many more species than had already been described by MARTIN **(Sangiran, Tjidjadjar** and **Simo),** the reliability of the established ratio of living species greatly depends upon the number of *Molluscs* occurring.

But even now it is easily understood that the ratio of 33% living species determined in **Tjilanang** from 189 species occurring, will not vary much when new discoveries are made in this spot, whilst new finds in **Pangka** — where only 18 species are known — will probably have much influence upon the ratio of 61% as now published.

It is a fact that in 1918 the ratio of living species in **Njalindung** was determined as 21.6% out of 97 species. It decreased to 18% after 95 more species had been discovered. In **Tjilanang** in 1918 the ratio was 34% out of 122 species. A decrease of only 1% resulted from the determining of 67 more species. On the other hand, as will be shown later on, in **Sangiran (Kali Tjemoro)** formerly a ratio of 33% living species was found from 21 species described by MARTIN, whereas the author was able to determine a ratio of 45% out of 76 species. This makes it necessary to exercise great caution in determining the age of the beds from the ratio of living species when too few Molluscs are present.

The following horizons, where great numbers of fossils occur, are to be considered as well fixed and as almost invariable in future with regard to the ratio of living species now established.

| | | | |
|---|---|---|---|
| Pliocene . . . . . . | **Sonde** | 53%, | 150 species known |
| lower part of Upper Miocene . . . . . . . | **Tjilanang** | 33%, | 189 „ „ |

| Formation | Occurrence | Total species of molluscs | % of living species |
|---|---|---|---|
| Quaternary | **Grissee** | 30 | 90 |
| | **Batavia to a depth of 6 M.** | 22 | 86 |
| Pliocene | **Kedung waru** [1] | 60 | 70 |
| | **Pangka** | 18 | 61 |
| | **Tjidjadjar** | 34 | 56 |
| | **Tjimantjeurih** | 45 | 55 |
| | **Waled** | 74 | 54 |
| | **Sonde** | 150 | 53 |
| | **Batavia from a depth of 74—92 M.** | 31 | 53 |
| | **Gombel** | 39 | 51 |
| | **Tjidjurai** | 64 | 51 |
| | **Tjikeusik** | 57 | 51 |
| Upper Miocene [2] | **W. of Parungponteng** | 33 | 45 |
| | **Tjiodeng** | 43 | 42 |
| | **Palabuanratu** | 26 | 38 |
| | **Ngembak, depth 60—70 M.** | 11 | 38 |
| | **Tjilanang** | 189 | 33 |
| | **Occurrence R. (Junghuhn)** | 42 | 31 |
| | **Tjilintung-Angsana** | 24 | 29 |
| | **Tjadasngampar** | 71 | 27 |
| Lower Miocene | **Njalindung** | 192 | 18 |
| | **Rembang** | 59 | 16,9 |
| | **West Progo mountains** | 110 | 8 |
| Upper Eocene | **Nanggulan** | 136 | 0 |

| | | | |
|---|---|---|---|
| upper part of Lower Miocene. . . . . . . . . . | **Njalindung** | 18%, | 192 species known |
| lower part of Lower Miocene . . . . . . . . . | **Westprogo-mountains** | 8%, | 110 „ „ |
| Upper Eocene | **Nanggulan** | 0%, | 142 „ „ |

[1]) from an unpublished report by MARTIN.

[2]) Some occurrences described by MARTIN as Upper Miocene are not mentioned here as the present author regards this determination as erroneous or doubtful [**Batavia** deepest horizons, **Kali Tjemoro** (= Lower Pliocene of **Sangiran**) and **Tambakbatu**].

It is evident that between these horizons great gaps exist, which are to be filled in by horizons represented by some of the other occurrences or by new discoveries. **Tjiodeng** (42%), for instance, has been stated by Martin to represent a horizon of the Upper Miocene younger than **Tjilanang** (33%), whereas no great disparity of age was supposed to exist between **Tjiodeng** and **Palabuanratu** (38%).

**Tjiodeng** and **Palabuanratu** seem to represent the upper part of the Upper Miocene. Considering the gap still existing between **Tjiodeng-Palabuanratu** (42—38%) and **Sonde** (53%) there is some possibility that other beds not yet described from Java will fit into it. Another and even much greater gap exists between **Sonde** (53%) and the Upper Quaternary beds of **Grissee** (90%) and **Batavia** (86%).

Martin also used another method to compare the age of a newly discovered occurrence of *Molluscs*. After having fixed a sort of division for the Neogene *Mollusc*-bearing beds he started to establish the vertical distribution of each species separately. The result is that, putting aside a large number of persistent species, there are several that apparently seem to be restricted to certain horizons only.

When investigating a newly discovered fossil *Mollusc*-fauna it will always happen that a number of these apparently restricted species have in reality a wider vertical distribution than had at first been supposed. In a Pliocene horizon, for instance, next to a number of Pliocene or Pliocene-recent species some species thought to be restricted to the Miocene will be found to appear. But the ratio of apparent Miocene species will be much lower than that of Pliocene forms, so that the Pliocene age will still be provable. On the other hand in an uncertain horizon the ratio of species thought to be restricted either to the Miocene or to the Pliocene will give sufficient evidence to determine the age of the beds. In this way the Miocene age of the **Tjiodeng** beds was concluded by Martin from the great number of species known to be restricted to the Miocene though the percentage of living species was just between that of the Pliocene **Sonde**-beds and that of the Miocene **Tjilanang**-beds.

Ultimately the additional knowledge gained regarding the vertical distribution of different species of *Molluscs* will lead to the establishment of guide fossils. The greater part of the species of *Molluscs*, however, will appear to be too persistent to be of any avail.

MARTIN (1931) succeeded in correlating some of the beds known from Burma and N.W. India.

| | | | |
|---|---|---|---|
| Pliocene: | **Sonde beds** | = **Gwádar**stage | **Mekran** |
| Upper Miocene: | **Odeng** beds | = **Talar**stage | N.W.India |
| Lower Miocene: | **Njalindung** beds | = Upper **Gáj** | **Gáj**(N. |
| | **Rembang** beds | = Lower **Gáj** | W.India) |
| Upper Eocene: | **Nanggulan** beds | = **Yaw**stage | (Burma). |

3. Fresh-water Molluscs.

The Pleistocene and partly the Pliocene being discovered to consist of terrestrial beds, a thorough investigation of the species of *fresh-water Molluscs* will lead to important results with regard to a demarcation of Pliocene and Quaternary beds.

The *fresh-water* beds however, are as yet insufficiently recorded and the determination of the *Molluscs* has been neglected, excepting those of the **Trinil** beds.

A comparative study of the *fresh-water Molluscs* of the Pliocene and Quaternary beds is to be considered as very urgent and should likewise embrace the phylogeneric evolution of some genera and species of *Melania* and *Paludina*.

A better knowledge of these Molluscs is of special interest with regard to those occurring in or alternating with an established marine fauna.

As *marine Molluscs* seem to fail in the Lower Quaternary, the possibility of a stratigraphic division of the Quaternary beds depends for a great deal upon the investigations regarding the *fresh water Molluscs*.

4. Vertebrates.

The determination of *vertebrates* by MARTIN, DUBOIS and the SELENKA EXPEDITION forms a basis for continued investigations.

It now having been established that the *vertebrate* series embrace Pliocene and Quaternary beds, great weight is to be ascribed to the principle that collections from different occurrences have to remain well separated. Further studies with regard to the phylogeneric evolution of *Elephantoids*, will, of course, be of great interest.

After the conclusion of STREMME — already mentioned in a former chapter in which the available material was fully discussed — the author refrains from stating a definite opinion. It seems that a great amount of work still remains to be done, not only in regard to the collecting of new material but also with regard to the description of existing collections.

As to the established species of vertebrates occurring in the Trinil and corresponding beds, the following list by MARTIN is still to be considered as representing our present-day knowledge.

LIST of vertebrates from the vertebrate beds at Patihajam, Sangiran, Trinil and Kendeng Mountans, according to MARTIN (1919).

| | | | |
|---|---|---|---|
| Pisces: | Selachii: | Carchariidae: | *Carcharias gangeticus* Müll. Henle. |
| | | | *Carcharias (Prionodon)* spec. indet. |
| | | Pristidae: | *Pristis* spec. indet. |
| | Teleostei: | Seluridae: | *Pimelodus* spec. indet. |
| | | | *Clarias batrachus* Linn. |
| | | Labyrinthici: | *Ophiocephalus* spec. indet. |
| | | | *Anabas microcephalus* Bleeker. |
| Reptilia: | Lepidosauria: | Varanidae: | *Varanus* spec. indet. |
| | ; | Colubridae: | *Colubridae* spec. indet. |
| | Testudinata: | Testudinidae: | *Batagur Siebenrocki* Jaekel. |
| | | | *Batagur signatus* Jaekel. |
| | | | *Hardella isoclina* Dubois. |
| | | Trionychidae: | *Trionyx trinilensis* Jaekel. |
| | | | *Chitra Selenkae* Jaekel |
| | | | *Chitra minor* Jaekel. |

| | | | |
|---|---|---|---|
| | Crocodilia: | Gavialidae: | *Gavialis bengawanicus* Dubois. |
| | | Crocodilidae: | *Crocodilus ossifragus* Dubois. |
| Mammalia: | Carnivora: | | *Mececyon trinilensis* Stremme. |
| | | | *Lutra palaeoleptonyx* Dubois. |
| | | | *Hyaena bathygnatha* Dubois. |
| | | | *Felis oxygnatha* Dubois. |
| | | | *Felis trinilensis* Dubois. |
| | | | *Felis microgale* Dubois |
| | | | *Feliopsis palaeojavanica* Stremme. |
| | Cetacea: | | *Sebaldius schlegeli* Flower. |
| | | | *Physatus antiquorum* Gray? |
| | Edentata: | | *Manis palaeojavanica* Dubois. |
| | Rodentia: | | *Hystrix* spec, indet. Stremme. |
| | Ungulata: | | *Tapirus pandanicus* Dubois. |
| | | | *Rhinoceros sivasondaicus*, Dubois. |
| | | | *Rhinoceros kendengindicus*, Dubois. |
| | | | *Sus brachygnathus* Dubois. |
| | | | *Sus macrognathus* Dubois. |
| | | | *Hippopotamus sivajavanicus*, Dubois. |
| | | | *Cervulus kendengensis* Stremme. |

*Cervus Lydekkeri* Martin.
*Cervus kendengensis* Dubois.
*Cervus palaeomendjangan*, Dubois.
*Duboisia Kroesenii* Dubois.
*Buffelus palaeokerabau* Dubois.
*Leptobos Groeneveldtii* Dubois.
*Leptobos dependicornis* Dubois.
*Bibos palaeosondaicus* Dubois.
*Bibos protocavifrons* Dubois.
*Mastodon* spec. indet [1]).
*Stegodon Airawana* Martin.
*Stegodon trigonocephalus* Martin.
*Elephas hysudrindicus* Dubois.
*Elephas* spec. indet.

Primates:
*Semnopithecus* spec. indet.
*Macacus nemestrinus* Linn.
*Macacus* spec. indet.
*Pithecanthropus erectus* Dubois.

---

[1]) Erroneously left out of this list by Martin.

# CHAPTER V

## SEA LEVEL VARIATIONS

1. Tertiary.

The signs of shifting of shoreline occurring in the Tertiary period are all to be considered as a sequel of orogenic movements.

Nevertheless a local origin of the movements is not always necessary; folding in distant regions may have caused great changes of land and sea and thereby may have influenced the level of the ocean.

Distinct transgressions of the sea are known to have occurred at the base of the Eocene and to have been repeated at the base of the Upper Miocene.

In another chapter it will be shown that the sea receded from anticlinal regions during the Upper Miocene.

The Pliocene is characterized by a shifting of the shoreline, perhaps even in a different sense, due to local movements.

The ultimate big Pliocene folding of the Tertiary beds resulted in the sea being removed from the folded and uplifted anticlinal regions, but it seems probable that part of the synclines remained submerged.

2. Quaternary.

*a*. Pleistocene.

In the *unstable* parts of the Archipelago rising or sinking movements of the land during the Quaternary may have caused some shifting of the shorelines.

For the *stable* parts, however, MOLENGRAAFF and WEBER (1919) in following CROLL (1875), PENCK (1882) and DALY (1910 and 1915), demonstrated that sea level variations were due to the freezing of enormous masses of ice in the snowcaps of Europe and America during the glaciation.

The *stable* part of the Malay Archipelago embraces Borneo, the East coast of Sumatra, the North coast of Java and the interjacent seas, viz. the South China Sea, Malacca Strait and the Java Sea.

Meanwhile the *unstable* parts of the Archipelago underwent the same changes, but the effect may have been lessened or increased by local movements.

Judging from various data regarding the present features of the sea bottom and the distribution of coral reefs, MOLENGRAAFF fixed the amount of the post-glacial rise of the sea level at about forty fathoms, but allowed for even higher figures.

He surmised that during the Pliocene the region now occupied by the Java Sea had been a low country or a group of islands. At the commencement of the Pleistocene the sea receded and the region formed a whole with Sumatra, Java and Borneo. The rivers joined and formed two big streams, one emptying into the South China Sea, the other into the Flores Sea, S. of Makassar Strait. At the end of the Pleistocene the sea level gradually rose and the lower regions were invaded by the present Java Sea and South China Sea.

In a second publication MOLENGRAAFF (1923) pointed out that the phenomenon is certainly more complicated than has just been described. Several ice periods were separated by at least three inter-glacial periods and the phenomenon of sea level variations repeatedly occurred.

UMBGROVE (1929) in studying the isobaths of the Java Sea, the South China Sea and Malacca Strait concluded that the depth of the former coastline was about 55 fathoms.

According to these conceptions one has to expect at least three inter-glacial and one post-glacial transgression.

But up to now in Java no traces have been found of the transgressions preceding the last one.

Nowhere in Java exposures have been described yet showing a marine fauna that might be taken for Pleistocene.

Evidently the highest sea level of the interglacial periods generally remained below the present sea level. The coastlines of earlier transgressions now being all covered by the sea, no traces of marine Pleistocene deposits are exposed in Java, excepting, perhaps, those regions where the Pleistocene beds have been slightly folded. A thorough investigation of the Quaternary layers met with in boreholes near the coast, may, of course, result in giving evidence regarding marine beds and sea level variations of the Pleistocene. But the fixing of the age of the beds will certainly prove to be extremely difficult, as a considerable number of *Molluscs* is necessary for a reliable determination.

*b.* Recent.

DALY (1920) surmised a recent sinking of the sea level since 2000—4000 years B.C. to an amount of 5—6 M., embracing all coasts of the world ocean, and ascribed it to a recent increase in glaciation.

UMBGROVE (1928) after first objecting to this theory with regard to the coral islands in the bay of Batavia strongly supported the conceptions of DALY in giving several proofs of regression in the islands in the bay of Batavia (1930), the South coast of Java near Tjilaoet eureun (1931) and the West coast of Sumatra (1931).

In Bangka VERBEEK described the occurrence of marine *Molluscs* in one of the tin mines. The present author established the existence of a subrecent coastline N. of Pangkal Pinang.

Though Java is considered to belong to the *unstable* parts of the Archipelago, it is quite improbable that all signs of a recent regression of the sea are to be ascribed to recent tilting movements. The traces of a slightly higher sea level in recent centuries are many.

The author investigated the rows of dunes occurring along the South coast of Java between Jogjakarta and Tjilatjap. They had already been described by VERBEEK and FENNEMA (1896) together with similar dunes in South Probolingo. Conformable formations occur in the bays of Poppoh and Patjitan. This proves the wide extension of the phenomenon.

The sand of the dunes contains a considerable quantity of magnetic iron ore and consequently the dunes were originally formed bordering the beach, as the magnetic iron ore is too heavy to have been removed by the wind. In examining the valleys between the rows of dunes the author ascertained that magnetic iron was indeed absent there.

There are several rows of dunes; the farthest inland represent the oldest coastline. This is wholly confirmed by the still unpublished archaeological discoveries made by the author in the region discussed.

The row of dunes farthest inland contains *early neolithic* objects mainly potshells of a type known from the deeper beds in the cave of Sampung in Madiun and from the caves of Dander in Bodjonegoro. The age of these cultural objects dates from at least 2000 B.C.

The coastline of this period in Jogjakarta is 1 Km. distant and in Tjilatjap about 3 Km. distant from the present coast. A great number of discoveries all restricted to this row of dunes was made in the interjacent region.

Another row of dunes much nearer to the coast at about $\frac{1}{3}$ of the distance of the former row contains pot shells of a wholly different character. It indicates the coastline of the *bronze period* in Java and dates from the beginning of the Christian aera.

Both discoveries prove a regression of the sea along the south coast of Java. Even farther West in Priangan the traces of this regression are visible in the cave-like excavations made by the sea in the tertiary coral limestones indicating an old coastline. (See photos 1 and 2).

By drilling a great many shallow boreholes the author established the regression of the sea in Madura, where several coastal plains are to be considered as old lagoons closed in by sand dunes and now raised above the present sea level.

In the Darmo aerodrome at Surabaja the author likewise discovered marine *Molluscs* at a shallow depth in horizontal beds occurring about 2—3 M. above sea level.

These correspond to the marine Quaternary beds described by MARTIN (1919) extending from Grissee northward to Bunga and westward to Lamongan 22 Km. from the present coast. They occur only slightly above the present sea level and contain 30 species of *Molluscs* of which 27, or 90%, are still living.

As even the remaining three forms are not proved to be extinct, MARTIN concluded a very young Quaternary age, which confirms the conclusions arrived at from the archaeological discoveries on the South coast of Java.

When, finally, the rapid extension of deltas along the North coast of Java is taken into consideration, it is evident that a recent regression of the sea is a feature not restricted to certain parts of Java but applying to the whole island. This statement makes it improbable that the origin is of a local character and, therefore, only variations of the sea level in the sense of DALY's hypothesis are more likely to have caused it.

1. Fossil surf-excavation in miocene coral limestone (S-coast of Priangan)
Photo Ir. F. T. MESDAG

2. Excavated rock and former abrasion-surface (S-coast of Priangan)
Photo Ir. F. T. MESDAG

# CHAPTER VI

## DESCRIPTION OF THE GEOLOGY OF REGIONS INVESTIGATED

### 1. **Baribis.**

*(With one map and two sections).*

*Situation*:

The region mentioned is situated in the western part of the residency of Cheribon, some Km. N, of Madjalengka and S. of the arterial road from Bandung to Cheribon.

The *orography* is characterized by a curved row of three parallel hill ranges partly encircling a central region of gently sloping lower hills out of which a few very steep hills arise.

*Stratigraphy*:

Lower Pliocene[1]).

*a. Dark gray argillaceous beds.*

The central region of low hills mainly consists of badly stratified, dark gray, marine argillaceous beds, containing but a few small-sized Globigerines and in some spots crystals of pyrite, which are often accompanied by gypsum derived from the weathering of the pyrite.

The argillaceous beds are strongly compressed and folded, especially in the vicinity of the encircling range of higher hills.

As no fossils have been found defining the age of the beds, there is no certainty as to what stratigraphic horizon they represent. The habitus of the beds shows great resemblance to the Lower Pliocene argillaceous beds N. of the Kendeng mountains and differs greatly from the Upper Miocene beds in that region.

The overlying strata contain molluscs of Pliocene age. As there

---

[1]) The demarcation between the Miocene and Pliocene will be discussed in the VIIth Chapter and the problem of the age of the Pleistocene beds will be treated in the VIIIth Chapter.

is no reason to surmize a reduced thickness of the Pliocene in this part of Java a Lower Pliocene age of the dark argilleous beds is not unlikely.

Middle Pliocene.

*b. Coral limestones.*

Adjoining the innermost of the three parallel bordering hill ranges near its southern extremity, there occurs a hill of coral limestone bearing the name of Pasir Oeroeg.

The limestone is to be considered as a coral reef of considerable thickness, but, as is to be expected, it soon wedges out in a N. and S. direction. In the higher levels the limestone passes into sandy marl.

Farther to the N., where the limestone fails, this marl bed corresponds to a tuff layer intercalated in argillaceous beds.

The limestone often contains masses of large coral colonies. Spines of cidaris abound.

In the coral limestone and the sandy marl several species of *Molluscs* occur. A list given by SIEMON (1929) contains only two determinable species of *Molluscs* and a great many undeterminable forms. Part of a collection of *Molluscs* gathered by the author in 1926 and 1930 has been examined with the following results:

Out of 31 species only 25 were found to be determinable.

Of these, 14 or 56% belong to still living species, which points to a Pliocene age **(Tjidjadjar** 56%, **Waled** 54%).

Besides 12 persistent forms there are 5 species up to now exclusively known from the Upper Miocene, whilst there are 8 species restricted to Pliocene or Pliocene and younger beds. The higher ratio of Pliocene species confirms the Pliocene age. but the occurrence of so many Miocene forms (20%) probably indicates a slightly greater age than the compared Middle Pliocene beds. Meanwhile 11 species including 4 of the abovementioned Miocene forms, likewise occur in the Upper Miocene **Tjilanang** beds.

Only 8 species including 7 persistant forms occur in the Lower Pliocene **Sangiran** beds. There is little resemblance to the Pliocene of **Tjidjadjar:** only 7 species correspond, but in the Pliocene of **Waled** 12 species, or 48%, are similar.

| *Molluscs* from the coral limestone and marl of Pasir Oeroeg, **Baribis**, Species determined by the author with the assistance of Dr. R. von Koenigswald | Miocene | Pliocene | Recent | Upper Miocene **Tjilanang** | Lower Pliocene **Sangiran** | Middle Pliocene **Tjidjadjar** | Middle Pliocene **Waled** |
|---|---|---|---|---|---|---|---|
| *Gasteropods*: | | | | | | | |
| Bulla ampulla Linn. . . . . . | + | + | + | + | | | + |
| Terebra indica Mart. . . . . | + | — | — | | | | |
| Conus socialis Mart. . . . . . | + | + | — | | | | |
| Surcula punctata Reeve . . . | — | + | + | | | | |
| Oliva funebralis Lam. . . . . | + | + | + | | | | + |
| Olivancillaria gibbosa Born, var. . . . . . . . . . . | + | — | — | + | | | |
| Marginella quinqueplicata Lam. . . . . . . . . . . | + | + | + | + | + | + | |
| Marginella quinqueplicata Lam. var. minor Mart. . . | + | + | + | + | + | + | + |
| Marginella dactylus Lam. . . | + | + | + | | + | + | + |
| Turricula taeniataeformis Mart | — | + | — | | + | | |
| Ranella subgranosa Beck . . | + | + | + | | + | | + |
| Cassis pila Reeve . . . . . . | ? | + | + | ? | + | | + |
| „ herklotsi Mart. . . . . | + | + | — | | | | |
| Dolium costatum Desh. . . . | + | + | + | | + | | + |
| „ hochstetteri Mart. . . | — | + | — | | | | + |
| Strombus isabella Lam. . . . | — | + | + | | | + | + |
| „ varingensis Mart. . | — | + | — | | | + | |
| Potamides hochstetteri Mart. . | + | — | — | + | | | |
| „ herklotsi Mart. . . | + | — | — | + | | | |
| Natica powisiana Recl.. . . . | + | + | + | + | | | + |
| „ globosa Chemn. . . . | + | + | + | + | + | + | |
| Solarium perspectivum Linn. . | + | + | + | + | | + | + |
| Trochus neglectus Mart. . . . | + | — | — | + | | | |
| *Lamellibranchiates* | | | | | | | |
| Ostrea baribisiana Mart. . . . | — | + | — | | | | |
| Clementia papyracea Gray . . | — | + | + | + | | | + |
| 25 species determined . . . . | 17+1 | 20 | 14 | 11+1 | 8 | 7 | 12 |

6 species not determined: Surcula, Mitra, Triton, Potamides, Circe, Tapes.

Taking all in all there is every reason to accept a Middle Pliocene age for the coral reef of **Baribis.** but the horizon is perhaps slightly lower than that of **Waled** and **Tjidjadjar.**

Upper Pliocene.

*c. White dacitic tuffs and conglomerates.*

The covering layers generally forming the southwestern slope of the first bordering hill range consist of white sandy dacitic tuffs often bedded and sometimes containing gravels and volcanic boulder conglomerates, with boulders of dacite and pumice stone.

Grains of sanidine abound and produce some resemblance to the Upper Miocene and Pliocene white tuffs of Bantam, Palembang and the East coast of Sumatra. The whole thickness of the beds does not exceed 15 M., but the horizon is very persisting. It does not contain fossils and bears the features of a river deposit.

*d. Andesitic conglomerates, sandstones and tuffs.*

The superincumbent layer is likewise to be characterized as a river deposit. Conglomerates equally alternate with tuffs and sandstones, but the grains and pebbles consist of andesitic material. The total thickness of the beds is about 50 M. The sandstones and conglomerates are generally weathered to a brownish colour.

No *Molluscs* occur, but part of a molar of *Stegodon* was collected from here.

*e Gray argillaceous marls and sandy marls.*

The series of valleys separating the first and the second hill ranges seems to have been caused by erosion in easily weathered dark gray marls with intercalations of sandy marls.

The marine character of the beds is shown by the occurrence of *Molluscs*.

The list of fossils is very short and contains only:

| | Miocene | Pliocene | Recent |
|---|---|---|---|
| Ostrea lingua Sow . . . . . . | + | + | — |
| Arca oblonga Phil. [1]) . . . . . | + | + | + |
| Placuna placenta Linn [1]) . . . | + | + | + |
| Turritella n. sp. [1]) . . . . . . | — | + | — |
| Ostrea sp. . . . . . . . . . | | | |
| Tapes sp. . . . . . . . . . . | | | |

The fossils occur in the intercalating sandy layers situated in

[1]) Abundantly.

about the centre of the valleys. The number of fossils is wholly insufficient to deduce the age of the beds.

There remains to be mentioned, however, the frequent occurrence of a remarkable new species of *Turritella*. This bears some resemblance to *Turritella simplex* Jenkins — with which it has been confused by SIEMON (1929) — known from the Upper Miocene, in showing the same solitary marked keel. The angle of the cone, however, is much less and gives the new species a more slender form. The older coils do not show the marked keel, which is replaced by several finer spirals somewhat resembling those of *Turritella bantamensis* and *Turritella terebra*.

In comparing a great many specimens of the recent living *Turritella terebra* collected by Dr. OOSTINGH, it appeared that this species shows a great variability. Several specimens possess some resemblance to the described new fossil species, but the solitary keel is less marked.

The author knows the new species of *Turritella* to occur in dark argillaceous beds on the banks of the river Tjibeët forming the boundary between the residency of West Priangan and Buitenzorg, and in the uppermost marine layers of the Pliocene of Tjidjurai.

Another remarkable fact is the frequent occurrence of *Arca oblonga* and *Placuna placenta*.

The fauna bears a character pointing to its having been reduced to a few marine species only. It resembles in a way the Upper Pliocene *impoverished marine fauna* of **Sangiran, Baringinan** and **Simo,** described hereafter.

Though SIEMON (1929) described several species of *Molluscs* from argillaceous beds at Baribis, there is no certainty that they belong to the horizon described, as there is a second horizon of argillaceous beds that likewise contains *Molluscs*.

*f. Andesitic conglomerates, sandstones and tuffs.*

The second hill range is formed by conglomerates, sandstones and tuffs wholly resembling those of the first range.

They represent a second horizon of river deposits in which *Vertebrate* remains occur.

*g. Gray argillaceous marls.*

The second series of valleys corresponds to a horizon of easily eroded marine argillaceous marls containing marine *Molluscs* consisting of the following species:

| | Miocene | Pliocene | Recent |
|---|---|---|---|
| Dosinia boettgeri Mart. . . . . | + | + | — |
| Limopsis venusta Mart. . . . | ? | + | — |
| Surcula nodifera Lamk . . . . | + | + | + |
| Siphonalia paradoxica Mart. . | + | + | — |
| Olivancillaria cheribonensis Mart. . . . . . . . . . . | — | + | — |
| Ostrea sp. . . . . . . . . . | | | |
| Meretrix sp. . . . . . . . . | | | |
| Cardita sp. . . . . . . . . . | | | |
| Arca sp. . . . . . . . . . . | | | |

The number of fossils is insufficient to determine the age of the beds.

*h. Conglomerates and sandstones.*

The third hill range is composed of conglomerates and sandstones alternating with argillaceous beds bearing all signs of having been deposited by rivers.

The total thickness of the Upper Pliocene beds described amounts to 500 M. Though the marine fossils do not give sufficient evidence regarding the age of the beds, in facies they somewhat resemble the higher horizons of the Pliocene beds of **Tjidjurai.**

In **Tjidjurai** fossil *Vertebrates* likewise occur in terrestrial deposits between marine series.

It is to be noted, however, that *Turritella djadjariensis,* so abundant in **Tjidjurai** and **Tjidjadjar,** does not occur in **Baribis,** whilst on the other hand the new species of *Turritella* found in **Baribis** is absent in **Tjidjadjar** and is represented by a few specimens only in the uppermost marine layers of **Tjidjurai.**

This points to a slightly younger age than the lower part of the fossiliferous Pliocene of **Tjidjurai.** Such is confirmed by the probability of a similar age as the Upper Pliocene layers of **Sangiran, Baringinan** and **Simo,** that likewise contain an *impoverished marine fauna.*

P l e i s t o c e n e.

*i. Volcanic boulder breccia.*

At the southern extremity the three parallel hill ranges are covered by a reddish brown, weathered, volcanic boulder breccia, which likewise occurs in a low hill east of the ranges.

Though unconformably covering the Pliocene beds, the volcanic boulder breccia shows signs of a slight upheaval caused by tilting movements.

This may point to a Pleistocene age of the boulder breccia.

H o l o c e n e.

*j. Alluvial deposits.*

The alluvial deposits are represented by younger products of Tjeremai volcano, forming the lower terraces of the rivers.

Of course in some instances the deposits are of various age. Since the period in which volcanic products had been deposited in the valley of Tjibatoe (Tjibioek), this river cut a deeper bed into the underlying dark argillaceous marls.

*Tectonic movements.*

The three parallel hill ranges form the steeply dipping eastern, northeastern and northern bordering flexure of an anticline, the central part of which consists of strongly folded and compressed argillaceous beds.

In the vicinity of Pasir Oeroeg it is probable that the folding movement even caused a slight upthrust, the argillaceous beds west of it giving evidence of having been compressed pell mell, whilst east of it the dip of the conglomeratic beds exceeds the vertical.

*Volcanism.*

The isolated steep hills occurring in the central part appear to be volcanic dykes. The injected bodies probable never reached the surface, They consist of very fine grained and compact dacite, somewhat resembling hypabyssal rocks.

The surrounding argillaceous beds show but few traces of contact metamorphism, this being restricted to an extremely thin zone of spotted clay.

2. **Tjidjadjar.**

M i d d l e P l i o c e n e.

A collection of *Molluscs* from Tjidjadjar has already been described by MARTIN (1919). SIEMON (1929) published 9 more species. The author is now able to add a considerable number of *Molluscs* to the existing lists.

Though no *Vertebrates* have been found, a better knowledge of the Pliocene beds with regard to the age of the vertebrate series is of great importance.

The *Molluscs* occur in several horizons, the uppermost of which crops out at the mouth of the river Tjidjadjar into the main river Tjiwaringin, whilst the rest occur upstream on the banks of the river Tjidjadjar.

| *Molluscs* from argillaceous beds alternating with sandy marls at **Tjidjadjar**, partly described by MARTIN and by SIEMON and augmented by a new collection determined by the author with the assistance of Dr. R. VON KOENIGSWALD | Miocene | Pliocene | recent | **Sangiran.** Lower Pliocene | **Tjidjurai.** Middle Pliocene | **Waled.** Middle Pliocene |
|---|---|---|---|---|---|---|
| *Gasteropods* | | | | | | |
| Terebra Cumingii Desh. [1] . . . | — | + | + | | | |
| Conus glaucus Linn. . . . . . . . | + | + | + | | | |
| Surcula nodifera Lamk. var. . . | + | + | + | + | + | + |
| „ varinginensis Mart. . . . | — | + | — | | | |
| Pleurotoma carinata Gray var. . | + | + | + | | + | + |
| Drillia neglecta Mart. [1] . . . . | + | — | — | + | | |
| Olivancillaria cheribonensis Mart. | — | + | — | | + | |
| „ subulata Mart. [2] . | + | + | + | | | + |
| Marginella quinqueplicata Lamk. [1] . . . . . . . . . . | + | + | + | + | + | |
| Marginella quinqueplicata Lamk. var. min. . . . . . . | + | + | + | + | + | + |
| Marginella dactylus Lamk. . . . | + | + | + | + | + | + |
| Voluta scapha Gmel. var. ponderosa Mart. . . . . . . . . . | + | + | — | + | + | + |
| Melongena cochlidum Linn. [1] . . | + | + | + | + | | |
| „ madjalengkensis Mart. | + | + | — | + | + | |
| „ rex. Mart. [2]. . . . . | — | + | — | | | |
| Siphonalia paradoxica Jenk. . . | + | + | — | | + | |
| Dipsaccus canaliculatus Schum. . | + | + | + | | | |
| „ gracilis Mart. . . . . | + | + | — | + | + | |
| Nassa Verbeeki [2] . . . . . . . . | + | + | — | + | + | |
| Murex lebacanus [2] . . . . . . . | — | + | — | | | |
| Eutritonium pileare Lamk. var. . | + | + | + | + | + | |
| „ cingulatum Lamk. . | — | + | + | | + | |
| Persona cancellina Desh. . . . . | — | + | + | | + | |

[1] From a new collection gathered by the author in 1927.
[2] From the list published by SIEMON 1929.

| *Molluscs* from argillaceous beds alternating with sandy marls at **Tjidjadjar,** partly described by MARTIN and by SIEMON and augmented by a new collection determined by the author with the assistance of Dr. R. von KOENIGSWALD | Miocene | Pliocene | recent | **Sangiran.** Lower Pliocene | **Tjidjurai.** Middle Pliocene | **Waled.** Middle Pliocene |
|---|---|---|---|---|---|---|
| Strombus isabella Lamk. [1] . . . | — | + | + | | | + |
| „ varinginensis Mart. . . | — | + | — | | | |
| Potamides palustris Linn. var. . | + | + | + | | | |
| „ djadjariensis Mart. . | — | + | — | | | |
| „ cheribonensis Mart. . | — | + | — | | + | |
| „ jenkinsi Mart. [2] . . . | + | + | — | + | | |
| Telescopium titan Mart. . . . . | ? | + | — | | + | |
| Vermetus javanus Mart. . . . . | + | + | — | + | + | + |
| Turritella djadjariensis Mart. . . | + | + | — | + | + | |
| „ javana Mart. [2] . . . | + | + | ? | | | |
| Solarium perspectivum Linn. . . | + | + | + | + | | + |
| Natica lineata Lamk. . . . . . | + | + | + | + | | + |
| „ globosa Chemn. . . . . | + | + | + | + | + | |
| Globulus vestiarum Lam. [2] . . . | — | + | + | | | |
| *Lamellibranchiates* | | | | | | |
| Ostrea disciformis Mart. . . . . | + | + | + | + | + | |
| Anomya boettgeri Mart. . . . . | — | + | — | + | | |
| Placuna placenta Linn. [1] . . . | + | + | + | + | + | |
| Arca bistrigata Dkr. . . . . . . | — | + | + | + | + | |
| „ djadjariensis Mart. . . . . | — | + | — | | | |
| „ tambacana Mart. . . . . . | + | + | — | + | + | + |
| „ rhombea Born. . . . . . . | — | + | + | | | |
| „ inaequivalvis Brug . . . . | — | + | + | + | + | + |
| „ tegalensis Mart. . . . . . | — | + | — | | | |
| „ granosa Linn. . . . . . . | ? | + | + | | + | |
| „ oblonga Phil. . . . . . . | + | + | + | | | |
| Dosinia boettgeri Mart. [1] [2] . . | + | + | — | + | + | + |
| Venus chlorotica Phil . . . . . | + | + | + | + | + | + |
| Tapes undulata Born. . . . . . | + | + | + | | + | |
| „ liratus [1] . . . . . . . . | — | + | + | | | |
| „ papilionaceus Linn. [2] . . | — | + | + | | | |
| Cyrena rustica Mart. [1] . . . . . | + | + | — | | + | |
| Diplodonta everwijni Mart. [1] . . | + | + | — | | + | |
| Cultellus cf. olivaceus Metc. . . | — | + | + | | | |
| Siliqua winteriana Dkr.. . . . . | — | + | + | | + | |
| Corbula socialis Mart. [1] . . . . | + | + | — | + | + | |
| Total species . . . . . . 59 | 34+2 | 58 | 32+1 | 26 | 32 | 14 |

Several species not determined.

[1]) From a new collection gathered by the author in 1927.
[2]) From the list published by SIEMON 1929.

The list numbers 59 species of *Molluscs* of which 33 or 56% belong to still living species. Several species, however, have not been determined yet, so there is some possibility of a lower percentage. MARTIN (1919) described only 34 species, 56% of which are still living.

MARTIN (1926) already pointed out the similarity to the Pliocene of **Tjidjurai** and the disparity to the Pliocene of **Waled.**

According to the new list 32 species or 54%, are conformable to **Tjidjurai,** against only 14 species, or 24%, to **Waled.**

MARTIN ascribed the disparity between **Tjidjurai** (+ **Tjidjadjar)** and **Waled** to a slight difference in age. One might as well explain this by the wholly different facies, as **Waled** possesses calcareous sandy beds and **Tjidjurai** shows a succession of alternating argillaceous marine beds and river deposits.

One should not forget, however, that facies may be caused by tectonic movements. This, on the other hand, signifies that a difference in facies might possibly be caused by a difference in age.

A comparison with the Lower Pliocene molluscs of **Sangiran** yields a number of 26 or 44% of corresponding species, which is less than **Tjidjurai.**

There being signs of the existence of Pliocene horizons *younger* than **Tjidjadjar, Tjidjurai, Waled** etc., the author proposes to call the latter Middle Pliocene.

### 3. Tjidjurai.

(*with one section through the Pliocene beds*).

Lower, Middle and Upper Pliocene.

The basal layers consist of argillaceous marls, almost void of fossils with the exception of Ostrea lingua occurring in some levels only.

The total thickness of the basal layers has not been determined, but it certainly exceeds several hundred meters.

The superincumbent marine *Mollusc*-bearing beds appear to alternate with sandy and conglomerate beds containing material of volcanic origin.

The conglomerates are to be considered as river deposits. Other beds consisting of black bituminous clays with fresh-water fossils show signs of having been deposited in marshes.

The fresh-water *Melania* have not been washed down into the marine series, as has been supposed by MARTIN (1926), but in reality possess an autochtone origin, as they are not worn off and occur in separate layers.

In one of the argillaceous beds containing *Melania* a femur of *Bos* was discovered by the author. A skull of *Stegodon*, stated to have been found loose in the river in the vicinity of this exposure, was shown to the author in 1926.

It seems that the *Vertebrates* are not confined to the fresh-water beds. The author is in possession of a specimen from this locality consisting of three vertebrae of *Bos*, covered with shells of *Balanus* and *Ostrea*.

The succession of marine and terrestrial beds is to be considered as a deltoid deposit.

The uppermost stratum is a river-deposited gravel bed, near the base containing rounded thick-shelled marine *Molluscs* showing much wear, which of course were derived from already denudated older marine beds. The immediately underlying marine bed contains the new species of *Turritella* that has been described from **Baribis.**

In **Tjidjurai** Quaternary volcanic deposits cover unconformably the Tertiary beds and prevent the investigation of possibly occurring higher levels than are shown in the exposed stratigraphic section.

4. **Waled.**

L o w e r  P l i o c e n e.

South of **Waled** or the canyon of **Menengteng,** the deeper levels consist of dark argillaceous beds, void of fossils. At some spots coarse to finer-grained tuffs occur, corresponding to eruptions of ash of a contemporaneous southern volcano.

With the exception of these tuff beds the stratification is generally very indistinct; still it seems that the beds are strongly folded. An exposure not far from the canyon on the right bank of the Sengarung river showed the following section:

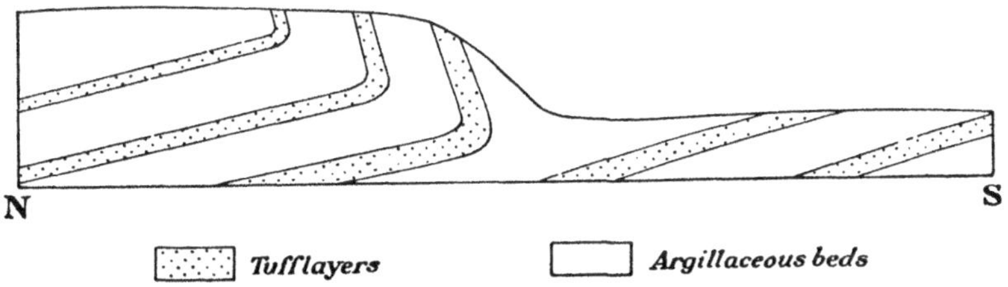

Middle Pliocene.

The beds exposed in the canyon are of quite a different character. They consist of sandy marls in thinner and coarser layers and show an apparent unconformity. The marine *Molluscs* occurring in the marls have been described by MARTIN (1919).

The dip of the beds varies from 55° to 35° N.E.

The strongly folded and compressed argillaceous beds probably represent the Lower Pliocene, while the mollusc-bearing beds are certainly Pliocene.

The manner of compression of the argillaceous beds shows the tendency of a strong upthrust movement from the S. and gives reason to accept the probability of an upthrust at the boundary between the argillaceous beds and the sandy marls.

The dips in the former are certainly not those of a normal flexure.

### 5. **Bumiaju.**

*Situation.*

The region containing fossil vertebrates is a hilly country West of the road from Tegal to Bumiaju and about 6 Km. N. of the latter village.

The rivers flowing in N.W. direction are all affluents of the Pemali River.

The region shows signs of a recent upheaval; erosion still has great effect and in the valleys landslips frequently occur. In the SE. part the folded Tertiary beds are covered by a plateau of weathered tuffs, originally belonging to the slopes of Slamat volcano: but it seems that this is likewise brought into a higher level and now juts out above the adjacent tuff slopes.

*Stratigraphy.*

According to a section of ZWIERZYCKI published by 'T HOEN (1930) two synclines are present, separated by an anticline with a core of Miocene beds. But TER HAAR (1929) showed in the axis of this anticline only argillaceous and sandy marls of the Middle Pliocene Turritella beds.

In the alleged southern syncline the dips vary greatly; the structure is not so simple as has been represented in the section published.

The report of the Java Geological Survey (1929) published the following succession of the beds:

| Age | Sediments | Fossils | Thickness in meters |
|---|---|---|---|
| Pliocene | 1. conglomeratic series with sandstone layers<br>2. tuffhorizon of grayish-white sandstones.<br>3. sandstone-conglomerate series with argillaceous sandstones, clay and marl, locally with beds of lignite and mostly bearing lime. | Scattered *vertebrate* remains<br><br>*Mastodon perimensis Stegodon airawana Hippopotamus.* Several *Melania* and *Corbicula* 69% [1]) living species; some marine *molluscs* and *foraminifera.* | 250<br>150<br>200 |
| | blueish-gray and greenish argillaceous marls, subordinate sandy marls.<br>Mollusc bed. | *Molluscs* with 56% [1]) recent species.<br>*Globigerina.* | 175 |
| | andesitic sandstones and conglomerates | In the upper part locally coral limestone. | 200-250 |
| Miocene | hard, — usually coarse basic andesitic breccia. Interjacent tuff series and horizon of pumice stone-breccia of 200 M. thickness 500 M. from the base; locally lava | | 1000 |
| | fine and coarse greenish andesitic sandstones and lime-bearing sandstones with interjacent marl beds, increasing in coarseness near the top.<br>A series of lightly coloured tufsandstone and grit with pumicestone forms a characteristic horizon. | | ? |
| | greenish spheroidical Globigerina marls bedded by thin layers of sandstone; nearing the base grayish-green marls with banked limestone containing many foraminifera; at the base unbedded speroidical argillaceous marls. | *Trybliolepidina rutteni* in several varieties.<br>*Cycloclypeus neglectus.*<br>*Cristellaria sp.*<br>*Operculina sp.*<br>*Operculinella sp.*<br>*Amphistegina sp.*<br>*Gypsina globulus.*<br>*Rotalia sp.*<br>*Lithothammium sp.* | ? |

[1]) according to new determinations mentioned in the following pages.

In the description given by the Java Geological Survey the following statements are added:

„The vertebrate-horizon of Kali Glagah lies conformably upon the *Turritella* bed.

The interjacent zone of several tens of meters consists of alternating marine beds and terrestrial andesitic conglomerates containing lumps of lignite corresponding to beds of 150 M. thickness in the **Bentarsari** basin, in which a bed of impure lignite occurs containing 50% water.

The lowest vertebrate beds are argillaceous marls and tuffoid sandy marls of andesitic material, containing *Paludina javana, Corbicula fluminea* and several species of *Melania* together with *Foraminifera*, viz. *Globigerina bulloides, Orbulina universa, Amphistegina lessonii, Pulvulina tumida, Cristellaria rotulata* etc.

The vertebrates that have been examined in Utrecht were derived from those beds. A Pliocene age is evident from the presence of *Mastodon perimensis* (?), *Stegodon airawana* and *Hippopotamus*.

In the *Turritella* zone, so called on account of the frequence of *Turritella djadjariensis*, 46 species of Molluscs were found to occur; 22 or 48% belonging to still living species. For the *Gasteropods*, the ratio is 15 : 31 and for the *Lamellibranchiates* 7 : 15. The fauna shows great resemblance to that of **Tjidjurai,** held to be Pliocene by MARTIN. The beds are intensively folded and faulted".

The list of marine *Molluscs* (p. 51) is derived from two combined collections from the *Turritella* bed in Bumiaju, sampled by C. A. DE JONGH and the author. Several species have not yet been determined.

It appears that of the 34 species determined, 19, or 56% are still living. This certainly points to a Pliocene age, which is confirmed by only 24 species know to occur in the Miocene, against 33 in the Pliocene.

A comparision with **Sonde** shows a great disparity to exist, but the resemblance to **Sangiran** and **Tjidjurai** is very obvious.

As **Sangiran** (hereafter to be described as Lower Pliocene) and **Tjidjurai** (Middle Pliocene) are different in age, the almost equal affinity to both faunas makes it rather difficult to decide

| Marine *Molluscs* from the Middle Pliocene of **Bumiaju** determined by author assisted by Dr. R. v. KOENIGSWALD | Miocene | Pliocene | Recent | Sangiran Lower Pliocene | Tjidjurai Middle Pliocene | Sonde Middle Pliocene |
|---|---|---|---|---|---|---|
| *Gasteropods* | | | | | | |
| Pleurotoma carinata Gray var. . . . . . . . . . . | + | + | + | + | + | + |
| Roualtia coronifera Mart. . | + | — | — | + | | |
| Trigonostoma crispata Sow . | — | + | + | | + | |
| Marginella quinqueplicata Lamk. var. min. . . . . | + | + | + | + | + | |
| Marginella dactylus Lamk. . | + | + | + | + | + | |
| Clavilithes Verbeeki Mart. . | + | + | — | | | + |
| Melongena cochlidum Linn. . | + | + | + | + | | |
| Siphonalia paradoxica Jenk. | + | + | — | | + | |
| Nassa Verbeeki Mart. . . . | + | + | — | + | + | + |
| Murex ejectus Mart. . . . . | — | + | — | + | + | |
| Purpura carinifera Lamk. . | — | + | + | | + | |
| „ costata Blain . . . | — | + | + | | + | |
| Ranella bitubercularis Lamk. | + | + | + | | + | + |
| Dolium costatum Desh. . . | + | + | + | + | | + |
| „ chinense Chemn. . . | + | + | + | + | | |
| Strombus isabella Lamk. . | — | + | + | | | + |
| Terebellum punctatum Chemn. . . . . . . . . | + | + | + | | | |
| Potamides palustris Linn. . | + | + | + | | | |
| Vermetus javanus Mart. . . | + | + | — | + | + | + |
| Turritella djadjariensis Mart. | + | + | — | + | + | |
| „ cf. sp. Baribis . . | — | + | — | | + | |
| Solarium perspectivum Linn. | + | + | + | + | + | + |
| Natica globosa Chemn. . . . | + | + | + | + | + | |
| *Lamellibranchiates* | | | | | | |
| Anomya boettgeri Mart. . . | — | + | — | + | | |
| Placuna placenta Linn. . . | + | + | + | + | + | |
| „ pseudoplacenta Mart. . . . . . . . . . | — | + | — | + | | |
| Chlamys senatorius Gmel.. . | + | + | + | + | + | + |
| Arca tambacana Mart. . . . | + | + | — | + | + | |
| „ biformis Mart. . . . . | + | + | — | + | | + |
| „ oblonga Phil. . . . . | + | + | + | | | |
| Dosinia Boettgeri Mart. . . | ′ + | + | — | + | + | |
| Corbula socialis Mart. . . . | + | + | — | + | + | |
| „ scaphoides Mart.. . | — | + | + | + | | |
| Pholas hercules Mart. . . . | — | + | — | — | + | |
| Total species 34 . . . . . | 24 | 33 | 19 | 22 | 21 | 10 |

from this point of view to what horizon the *Turritella* bed of **Bumiaju** belongs.

However, considering the rather high percentage of living species there is more reason to accept a Middle Pliocene age.

| Fresh-water molluscs from **Bumiaju** beds determined by Prof. Dr. H. GERTH | Miocene | Pliocene | Pleistocene | Recent | Sangiran (Upper Pliocene) | Trinil (Pleistocene) |
|---|---|---|---|---|---|---|
| Corbicula fluminea Müll. . . . . . . |  | + | + | + |  |  |
| Unio productus Mouss. . . . . . . . |  | — | + | + |  | + |
| Melania testudinaria v. d. Busch . . |  | + | + | + | + | + |
| tuberculata Müller . . . . . | + | + | + | + |  |  |
| tjemoroensis Mart. . . . . . |  | + | — | — | + |  |
| infracostata Mouss. . . . . | — | — | + | + |  |  |
| Woodwardi Mart. . . . . . |  | + | — | — |  |  |
| madiunensis Mart. . . . . . |  | + | — | — |  |  |
| porcata . . . . . . . . . . |  | — | — | + |  |  |
| gloriosa . . . . . . . . . . |  | — | — | + |  |  |
| tornatella Lea. . . . . . . | + | + | + | + |  |  |
| soloënsis Mart. . . . . . . |  | + | — | — | + |  |
| Paludina javana v. d. Busch . . . . |  | + | + | + | + | + |
| Total species 13 . . . . . . . | 2 | 9 | 7 | 9 | 4 | 3 |

Of 13 determined species 9, or 69%, are recent living.

This percentage points to an Upper Pliocene age of the freshwater beds.

4 Species correspond to the fauna of the Upper Pliocene freshwater series of **Sangiran,** and only 3 to the Pleistocene fauna of **Trinil.**

Moreover the beds are steeply folded. As up to now Quaternary beds with a steep dip are unknown, a Tertiary age is most likely.

### 6. Mt. Patih-Ajam.

(*with geological map and one section*).

*Situation.*

The region investigated lies SSE. of Muriah volcano and — seen from the south — gives the impression of a strongly eroded volcano, of which Mt. Patih-ajam forms the highest summit. In reality the nature of the region is quite different. It is an eroded anticlinal dome, but a horizon of volcanic breccia protected the central part of the dome in such a way that it still forms the highest summits. To the N. the uppermost beds of conglomerates and tuffs pass into the tuff mantle of Mt. Muriah, from which they are separated by a feeble interjacent topographic saddle.

*Stratigraphy*:

Middle Pliocene?

*a. Marine argillaceous beds.*

The deepest levels are exposed in the valley of R. Wuluk and near campong Djambe. This village lies in a circular valley, closed in on all sides, except to the East, by steep walls of breccia and tuff.

In the lowest part of the valley argillaceous beds containing marine *Molluscs* are exposed. Some pebbles of limestone occur containing boreholes of *Molluscs*.

The number of *Molluscs*, however, is far too few to deduce the age of the beds. The argillaceous beds are badly stratified but show signs of having been much compressed. They are of a darkgray colour and in habitus much resemble the Pliocene argillaceous beds known from this part of Java. As the Miocene beds are of a widely divergent character, a Miocene age is not likely.

Upper Pliocene.

*b. Argillaceous lacustrine beds.*

The superincumbent stratigraphic horizon exposed in the valley of Djambe, consists of gray tuffoid fresh-water beds with small pebbles of andesite, greatly resembling the freshwater argillaceous beds of **Bumiaju** and **Sangiran.** No fossils occur. As in the valley of R. Wuluk and in the eastern part of Djambe valley this horizon was not exposed, it was represented in the map and the section as being of only local occurrence.

*c. Volcanic boulder breccia.*

The beds that protected the central part of the anticline against erosion mainly consist of volcanic breccia. On the Eastern slopes the breccia is less coarse and weaker, which favoured erosion. On the Southern slopes in the valley of K. Goro intercalations occur resembling lava beds.

The age of the breccia is unknown. In some spots *Melania* occur.

The volcanic breccia being covered by similar vertebrate beds, it is supposed by the author to represent the same horizon as the breccias of S. Kendeng mountains in the vicinity of Mt. Pandan.

Quaternary.

*d. Sandstones, conglomerates and clays.*

The next series of beds consist of much alternating fresh-water conglomerates, sandstones and clays.

They contain *Paludina javanica v. d. Busch,* diff. sp. of *Melania* and a great many remains of *Vertebrates.* A thin bed of fresh-water limestone mainly consisting of shells of *Paludina javanica* occurs N. of Mt. Slumprit. There is hardly any doubt that the beds represent the same horizon as the **Trinil** beds. The habitus of the deposits is similar to that of the main vertebrate horizon S. of Kendeng mountains and the fauna corresponds to that of **Trinil.**

MARTIN (8, 9) determined: *Stegodon airawana* Mart., *Stegodon trigonocephalus* Mart. and *Cervus Lydekkeri* Mart.

A collection by the author in 1926—1927 contains remains of *Cervus, Palaeokerabau, Sus* and of *turtles* and *fishes.*

Special mention must be made of part of a molar described by MARTIN (1888) as *Mastodon* sp., Though the occurrence of *Mastodon* in Quaternary beds was afterwards (1919) omitted in the list of *Vertebrates,* the writer is authorized by Prof. MARTIN to state that this omission was erroneous. The specimen much resembles a similar fragment collected by the author in **Sangiran**[1]).

*Vertebrate* remains were found in the whole circumference of Mt. Patih-ajam, but they occur most abundantly near Mt. Slumprit and W. of it.

*Melania* abound in a clayey bed on the Western slopes of Mt. Slumprit.

*e. Volcanic tuffs and conglomerates.*

The uppermost layers consist of volcanic tuffs and intercalating conglomerates.

In habitus and dip they show some resemblance to the **Notopuro** beds that cover the **Trinil** beds S. of the Kendeng mountains in the vicinity of Mt. Pandan.

1) Quite recently Prof. DUBOIS showed to the author a similar part of a molar collected from S. Kendeng Mountains, but to his opinion there is also some possibility of its being an atavistical deviation of a *Stegodon* molar.

*Tectonic structure.*

The region investigated appears to form an anticlinal dome attacked by erosion on the western, southern and eastern slopes. The dip of the marine beds near Djambe, which in all probability are Middle Pliocene, is sometimes very steep. This gave reason to accept an unconformity between the marine and the overlying fresh-water beds, though is must be admitted that it is also conceivable that either younger folding movements or pressure caused by the weight of the surrounding high breccia hills favoured the pressing out of the plastic argillaceous matter of the underlying marine beds. For this reason the unconformity is represented as still problematic and the steep dips mentioned have not been shown on the map.

In the volcanic breccia the dip varies from 8 to 23°. The vertebrate beds show a dip varying from 5—20°, and in the uppermost volcanic tuffs and conglomerates the dip is still less.

7. **Sangiran.**

*(with one map and one section).*

*Situation.*

The anticlinal dome of Sangiran is situated in the centre of the plain of Surakarta about 3 Km. E. of Kalioso and 10 Km. N. of Surakarta. The orography of the dome is best described as a circular valley. Seen from the central region, the valley seems to be surrounded by abruptly sloping hills, Mt. Kritjian possessing the highest summit (180 M). But these hills, though showing steep sides towards the valley, are only gently sloping outwards conformable to the dip of the beds. It is evident, that before erosion exercised its effect, the dome must have looked like a broad, gently sloping hill. This in way is the impression one still gets, when viewing it from afar.

Some Km. NW. and S. of Sangiran there are two other broad dome shaped hills, probably identically built, which give an idea of the original form of the Sangiran dome before it was attacked by erosion. The structure of these hills is not visible, of course, as the uppermost layers are still intact.

Erosion evidently had stronger effect upon the Sangiran dome in consequence of the stronger vaultlike upheaval, and the fact that the structure is crossed by the River Tjemoro.

Almost all rivers descending from the slopes of the Merbabu volcano flow northward in the direction of Semarang or choose a roundabout way southward before joining the Solo River. The Tjemoro is the only river that breaks straight through the row of dome-like hills.

Therefore there is some possibility of the Tjemoro river being a *superimposed* river, dating from the time previous to the folding of the Sangiran dome.

On the other hand erosion once having reached the softer argillaceous beds in the centre of the dome, a small affluent of the Solo river might soon have cut backwards into the river systems of Mt. Merbabu.

*Stratigraphy.*

Lower Pliocene.

a. *Blueish-gray clay.*

The oldest strata exposed in the centre of the dome are badly stratified, blueish-gray, marine argillaceous beds of unknown thickness. Fossils are scarce and belong to the following species:

*Galene obscure Edw.*

*Dolium zonatum Green.*

*Pholas hercules Mart.*

*Corbis sp.*

The *Crustacea* often occur in the core of lime concretions.

Being too few in number, the fossils do not help to solve the problem of the age of these beds, which has to be deduced from the conformably covering layers.

b. *Turritella beds (argillaceous sand.).*

The name of *Turritella* beds was applied to argillaceous sands alternating with blueish-gray marls and clays. The visible thickness of these beds is not likely to exceed 30 M. There are signs of the superincumbent bed of *Balanus*-limestone having been deposited unconformably.

The dip of the beds is sometimes rather steep, but this may possibly have been caused by later movements.

The fauna of marine *Molluscs* is rather rich in species and shows features of having lived in a muddy and shallow sea. *Corals*, represented only by two (not determined) species, are very scarce.

| Marine molluscs of the Turritella bed at Sangiran, after MARTIN (1919), considerably augmented by a new collection determined by the author with the assistance of Dr. R. VON KOENIGSWALD | Miocene | Pliocene | Recent | Tjilanang Upper Miocene | Bareng Middle Pliocene | Sonde Middle Pliocene |
|---|---|---|---|---|---|---|
| *Gasteropods* | | | | | | |
| Terebra tjilonganensis Mart. [1] | + | — | — | | | |
| „ pamotanensis Mart. [1] | + | — | — | | | |
| Conus vimineus Reeve [1] | — | + | + | | + | + |
| Pleurotoma carinata Gray var.[1] | + | + | + | | + | + |
| „ gendinganensis Mart. [1] | + | + | — | + | + | + |
| Drillia neglecta Mart. [1] | + | — | — | + | | |
| „ suturalis Gray [1] | + | + | + | | | |
| „ flavidula Lamk. [1] | — | + | + | | + | + |
| „ tjemoroënsis Mart. [2] | ? | ? | — | | | |
| „ sangiranensis Mart. [2] | ? | ? | — | | | |
| Roualtia coronifera Mart. [1] | + | — | — | + | | |
| Marginella quinqueplicata Lamk. [1] [2] | + | + | + | + | + | + |
| Marginella quinqueplicata var. min. Mart. [1] [2] | + | + | + | + | + | [4] |
| Marginella dactylus Lamk. [1] | + | + | + | | + | [3] |
| „ sangiranensis Mart. [2] | ? | ? | — | | | |
| Voluta scapha Gmel. var. ponderosa Mart. [1] | + | + | + | | | [3] [4] |
| Mitra circula Kien [1] | — | + | + | | | |
| Turricula taeniataeformis Mart. [1] | — | + | — | | | |
| Fusus menengtenganus Mart. [1] | — | + | — | | | |
| Cominella sangiranensis Mart. [1] [2] | — | + | — | | + | [3] |
| Clavilithes tjidamarensis Mart. [1] | + | — | — | | | |
| Melongena cochlidum Linn. [1] | + | + | + | + | + | |
| Dipsaccus pangkaënsis Mart. [1] [2] | + | + | — | | + | |
| „ gracilis Mart. [1] | + | + | — | + | + | |
| Nassa Verbeeki Mart. [1] [2] | + | + | — | | + | + |
| Murex Verbeeki Mart. [1] | + | + | — | + | + | + |
| „ ejectus Mart. [1] | — | + | — | | + | |
| Purpura depressa Mart. [1] | + | — | — | + | | |
| Eutritonium pileare Linn.[1] | + | + | + | + | + | |
| „ tjilonganensis Mart. [1] | + | — | — | | | |
| „ pseudopyrum Mart. [1] | + | + | — | + | + | + |
| Persona reticulata Linn. [1] | + | + | + | + | | + |
| Hindsia tambacana Mart. [2] | ? | + | — | | | + |
| „ tjemoroënsis Mart. [2] | — | + | — | | + | |
| „ gendinganensis Mart. [1] | — | + | — | | | + |
| Ranella subgranosa Beck. [1] | + | + | + | | + | + |
| „ nobilis Reeve [1] | + | + | + | + | + | + |
| „ pulchra Gray [1] | + | + | + | | + | |
| Cassis pila Reeve [1] | ? | + | + | ? | + | + |
| Dolium costatum Desh [1] | + | + | + | | + | + |
| „ chinense Chemn. [2] | + | + | + | + | | |
| Vermetus javanus Mart. [1] | + | + | + | + | + | + |

[1] Collected by author in 1927—1930.
[2] From the list by MARTIN.
[3] Collected by the author out of identical bed at Bandjarbangi, Sukun and Soko.
[4] Mentioned by MARTIN—ICKE from Padasmalang and Pengkol.

| Marine molluscs of the Turritella bed at Sangiran after MARTIN (1919), considerably augmented by a new collection determined by the author with the assistance of Dr. R. VON KOENIGSWALD | Miocene | Pliocene | Recent | Tjilanang Upper Miocene | Bareng Middle Pliocene | Sonde Middle Pliocene |
|---|---|---|---|---|---|---|
| Turritella bantamensis var. talahabensis Mart. [1] [2] . . . . . | + | + | — | | + | |
| Turritella cingulifera Sow [2] . . | + | + | + | | | [4] |
| „ javana Mart. [1] . . . | + | + | — | + | + | |
| „ djadjariensis Mart. [2] | + | + | — | + | + | |
| Solarium perspectivum Linn. [1]. | + | + | + | + | + | + |
| Natica lineata Mart. [2]. . . . . | + | + | + | | + | |
| „ marochiensis Mart. [2] . . | + | + | + | + | | |
| „ globosa Chemn. [1] . . . | + | + | + | + | + | |
| „ powisiana Mart. [1] . . . | + | + | | + | + | + |
| „ rufa Born. [2]. . . . . . | + | + | + | + | + | + |
| „ vitellus Linn. [2] . . . . | + | + | + | | | |
| Sigaretus papilla Chemn. [1]. . . | + | + | + | + | | + |
| *Lamellibranchiates* | | | | | | |
| Ostrea lingua Sow. [1] . . . . . | + | + | — | + | + | [3] |
| „ hyotis Linn. [1] . . . . . | + | + | + | + | | + |
| „ promensis Noetling [1] . . | + | + | — | + | | |
| „ disciformis Mart. [2] . . . | + | + | — | + | + | + |
| „ folium Gmel. [1]. . . . . | — | + | + | | | + |
| Anomya boettgeri Mart. [1] . . . | — | + | — | | + | |
| Placuna placenta Linn. [1] . . . | + | + | + | | + | [3] |
| „ pseudoplacenta Mart. [1] | — | + | — | | + | |
| Amussium hulshofi Mart. [1] . . | + | — | — | | + | |
| Chlamys senatorius Gmel. [1] . . | + | + | + | + | + | + |
| „ kebolintangensis Mart. [1] . . . . . . . | — | + | — | | + | + |
| Arca biformis Mart. [1] . . . . . | ? | + | — | | | + |
| „ bistrigata Dunker [1] . . . | — | + | — | | + | |
| „ bataviana Mart. [2] . . . . | ? | + | — | | | |
| „ tambacana Mart. [1]. . . . | ? | + | — | | + | |
| Limopsis venusta Mart. [1] . . . | ? | + | — | | + | ? |
| Cardita elegantula Desh [1] . . . | + | — | — | | | |
| Crassatella radiata Sow [1] . . . | — | + | + | | + | |
| Dosinia Boettgeri Mart. [1] . . . | + | + | | + | + | |
| Venus chlorotica Phil. [1] . . . . | + | + | + | + | + | |
| Corbula socialis Mart. [1] [2] . . . | + | + | — | + | + | [3] |
| „ scaphoides Hinds . . . | — | + | + | | + | |
| Total species 76 . . . . . | 51 | 64 | 34 | 31+1 | 47 | 26+8 |
| *Scaphopodes* | | | | | | |
| Dentalium Junghuhni Mart. . . | + | + | — | | + | |
| „ erectum Sow . . . . | ? | — | — | | | |

Species not determined:

Conus, Turricula, Siphonalia, Rimella, Buccinum, Arca 1, Arca 2, Pseudamussium (total 8).

Melongena madjalengkensis Mart. has been deleted from the list as this fossil was in all probability derived from a higher marine horizon.

---

[1]) Collected by the author in 1927—1930.

[2]) Form the list by MARTIN.

[3]) Collected by the author out of identical bed at Bandjarbangi, Sukun and Soko.

[4]) Mentioned by MARTIN—ICKE from Padasmalang and Pengkol.

Disregarding the three species occurring exclusively in **Sangiran,** those of uncertain vertical distribution and those that are as yet undetermined, there appear to be 76 species, 34 of which, or 45%, are still living.

This percentage is much higher than that of 33% formerly deduced by MARTIN (1919) out of only 21 species.

The percentage of 45% shows that Sangiran is certainly younger than **Tjilanang** (33% living species) and older than **Bareng** (53%) or **Sonde** (53%). It approaches the percentage of **Tjiodeng** (42%) and **Parung ponteng** (45%).

MARTIN (1919) pointed out that the fauna of **Tjiodeng** shows more affinity to those of the typical younger Miocene and proved **Parung ponteng** to be still Miocene, as most of the species correspond to **Tjadasngampar,** where only 27% living species occur.

On the other hand the author is of opinion that **Sangiran** possesses the features of a Pliocene fauna. Only 51 species are known from the Miocene, against 64 species known to occur in the Pliocene. The number of species which hitherto were considered as being confined to the Miocene is only 9, whilst 18 species occur that up to now were exclusively known from Pliocene or Pliocene and younger beds. This proves that the fauna shows more affinity to the Pliocene than to the Upper Miocene.

Moreover the facies does not resemble at all that of the uppermost Miocene beds in the Kendeng mountains N. of Sangiran, nor does it correspond to that of the Miocene beds in the Southern mountains of Java. It is evident that the fauna of **Sangiran** fills a gap between the Upper Miocene **Tjiodeng** beds and the Pliocene **Bareng** or **Sonde** beds. The author therefore proposes to designate the fauna of the *Turritella* bed of **Sangiran** as Lower Pliocene.

A comparison with the species occurring in the Upper Miocene of **Tjiodeng** and **Palabuanratu** showed that only 4 and 3 species respectively are identical. This also proves that the fauna of **Sangiran** can never be of the same age.

Out of 76 *Molluscs* 47 species of **Sangiran** or 62% correspond to **Bareng,** against only 31 species or 41% to **Tjilanang** and $26 + 8 = 34$ species or 45% to **Sonde.**

The discovery of a Lower Pliocene fauna shows that **Sonde, Bareng** etc. are at present to be considered as Middle Pliocene.

It is to be noted, that the fauna of **Sangiran (K. Tjemoro)** was formerly erroneously described as Miocene and, therefore, a number of fossils have been described by MARTIN as occurring in the Miocene, whereas in reality their vertical distribution is more restricted.

The facies of the argillaceous beds underlying the *Turritella* bed of **Sangiran** much resembles that of the dark argillaceous beds underlying the **Bareng** fossiliferous beds. As the fauna of **Sangiran** described is certainly older than that of **Bareng,** it appears that the *Turritella* bed of **Sangiran** is to be considered as forming a fossiliferous horizon of the dark argillaceous beds of Lower Pliocene age.

Middle Pliocene.

*c. Balanus limestone.*

A remarkable limestone mainly consisting of shells of *Balanus* and of small *Foraminifera,* seems to transgress unconformably over the *Turritella* beds. In **Sangiran** this unconformity was deduced from the much greater dip of the underlying beds. However, as there might be another explanation for this phenomenon, this fact is not quite so convincing. In the region of **Kaliuter** next described, however, there is an obvious unconformity between the *Balanus* limestone and the underlying marine argillaceous beds and it may therefore be assumed to exist also in **Sangiran.**

The *Foraminifera* occurring in the Balanus limestone partly have been determined by Dr. VAN DER VLERK.

*Biloculina anomala Schlumb.*
*Rotalia Beccarii Linn.*
*Elphidium (Polystomella) craticulata Ficht. and Moll.*
*Truncatulina sp.*
*Amphistegina sp.*
*Operculina sp.*
*Miliolina sp.*

The author noticed that the *Balanus* limestone forms a remarkable persistant horizon at the base of the Middle Pliocene in the region between Surakarta and Surabaja. In the **Sangiran** dome it occurs almost over the whole circumference, the S. excepted. The thickness of the bed varies from 0,5 to 1,5 M. It has been slightly exaggerated in the section.

The *Balanus* limestone was discovered again in **Kaliuter, Baringinan,** in several spots S. of the Kendeng mountains in the vicinity of Mt. Pandan and in the Kendeng mountains N. of Djombang. As to the occurrence in **Kaliuter** and **Baringinan,** the corresponding succession of the beds leaves no doubt as to its representing the identical horizon. In the other spots mentioned the corresponding age follows from the description and will be discussed later on.

One may expect that the factors that caused the formation of this remarkable breccia of *Balanus* shells did not often coincide. It seems to the author that the *Balanus* limestone is of great importance in correlating Pliocene beds in the Eastern half of Java.

The fact of its occurring in Mt. Butak and Tjabehan at the base of the Sonde beds defines the age as Middle Pliocene, which would correspond to the position in **Sangiran** where it overlies the Lower Pliocene *Turritella* beds.

Upper Pliocene.

*d. Corbicula bed.*

The *Balanus* limestone is immediately covered by a fresh-water bed consisting of a yellowish tuff full of shells of *Corbicula exporrecta Mart.* MARTIN (1883—1887) described this fossil from **Ngembak,** where the subsoil is proved to be Miocene by the occurrence of molluscs. He admitted, however, that this species of *Corbicula* belongs to a number of fossils that have possibly been collected from surface diggings. As by the description of the author it is now known that in **Ngembak** only slightly dipping *Vertebrate* beds unconformably cover folded Miocene beds, it is evident that *Corbicula exporrecta* most probably occurs in fresh-water beds at **Ngembak,** of which, in the author's opinion, a Pleistocene age is most likely.

In the southern part of the **Sangiran** dome the *Corbicula* bed is found to cover directly the *Turritella* beds, as the *Balanus* limestone is lacking.

In comparing the stratigraphic sequence of the beds in **Baringinan** and **Sangiran** it appears that in **Sangiran** a hiatus exists between the *Balanus* limestone and the *Corbicula* bed. In **Baringinan** marine argillaceous beds appear to occur between the *Balanus* limestone and the *Corbicula* horizon. This leads to the assumption of a second unconformity being probably present in **Sangiran.**

The *Corbicula* bed is rather thin on the southern slopes of the anticlinal dome of **Sangiran,** but it increases in thickness to several meters in the opposite slopes. Its appearance is the first token of a fresh-water lake existing in the Northern part of the plain of Surakarta during the period posterior to the forming of the **Sonde** beds. This lake extended far northward to **Baringinan** and remained in existance during a long period. As will be shown afterwards, it was once again invaded by the sea for a very short time.

*e. Volcanic breccia.*

The tuff of the *Corbicula* bed already indicated an increase of volcanism. The next layer is a true volcanic boulder breccia or „lahar" most probably originating from the eastern slopes of Mt. Lawu-Kukusan.

In the northern part of the **Sangiran** dome the boulder breccia wedges out to a mere 3 dm, whilst over the rest of the circumference the thickness amounts to 30 m. and more.

In the breccia the author was able to determine several freshwater fossils occurring together with marine *Molluscs.*

Fresh-water: *Corbicula exporrecta Mart.*
*Paludina javana v. d. Busch.*
*Melania sp.*
*Crocodilus sp.*

Marine: *Marginella quinqueplicata var. minor Lam.*
*Cominella sangiranensis Mart.*
*Vermetus javanus Mart.*
*Melongena cochlidum Linn.*
*Arca tambacana Mart.*
*Arca oblonga Phil.*
*Corbula socialis Mart.*
*Several sp. not determined.*

Notwithstanding the occurrence of marine fossils, the boulder breccia is not a marine deposit.

The phenomenon — frequently occurring in Java — indicated by the Javanese name of „lahar" forms a common element in the composition of the slopes of volcanoes.

After an eruption, volcanic ashes, heaped around the cone of the volcano, are swept by a big rain storm into the steeply descending valleys of the rivers.

The mud-laden water, having a high specific gravity and a great velocity, picks up all boulders from the river bed and undermines the steep slopes of the valleys, thereby adding new material to the already swollen rivers.

The swiftly descending „lahar" — in a way comparable to the „Muhrgang" in Switzerland — with a height often attaining several dozens of meters, arrives in the lower plains, where it floods the banks of the rivers and spreads out, destroying everything it meets, whilst often big tracts of land are covered by it.

As its velocity soon decreases where the slope is less, the water can no longer carry the same amount of solid matter. Often the content of water is even so small that it seems to have disappeared as soon as the „lahar" comes to a stop.

The „lahar" deposits are easily recognizable by their wholly unstratified and unassorted habitus. The material is mostly of volcanic origin and consists of tuff in large quantities mixed with sand and gravel and boulders of andesite of sometimes amazing dimensions.

Where the „lahar" passes over sediments of different origin, part of these may be easily picked up, especially as the eroding force is exceedingly great.

In this way it is to be supposed that the described volcanic qoulder breccia of **Sangiran** contains marine fossils from the underlying *Turritella* bed and fresh-water fossils from the *Corbicula* bed and possibly from other unknown fresh-water beds.

Lumps and fragments of *Turritella* sandstone and *Corbicula* tuff containing the fossils described and derived from the corresponding beds are certainly inclosed in the breccia.

It is to be noted that also *Vertebrate* remains occur, belonging to some species of *Crocodilus* (collected by the author and determined by Dr. R. VON KOENIGSWALD).

*f. Black clay.*

The main sediment of the fresh-water lake consists of a black clay, in which various other layers are intercalated. The clay is best described as a fine mud deposited in the centre of the lake.

The black clay contains fresh-water Molluscs in various quantities.

The lowest horizon is marked by:
*Corbicula exporrecta,*
*Paludina javana* and
*Unio sp.*

A higher horizon immediately overlying the first *Diatom* bed contains the same fauna, to which several species of *Melania* are added.

An identical fauna is found on top of the second *Diatom* bed. This horizon also contains *Vertebrate* remains.

The next horizon is a veritable shell bed of 1 M. thickness and contains *Corbicula exporrecta* together with *Isidora sp.* and a great many species of *Melania.* One species of *Melania* is remarkable by its particular form and in **Sangiran** seems to be confined to this bed only. It shows some affinity to *Melania testudinaria,* but differs widely in as much as it shows a marked solitary keel.

One specimen of *Melania sp.,* to all appearance fully resembling the form discussed, has been found together with *Paludina javana* and *Isidora* sp. in the Middle Pliocene marine fauna of **Sumberringen 1,** containing 59% living species.

In the upper layers of black clay of **Sangiran** several other horizons of fresh-water *Molluscs* occur, mainly containing *Paludina javana, Corbicula exporrecta* and a few species of *Melania* together with small *Vertebrate* remains. The highest beds, however, are void of fossils and only contain lime concretions.

MARTIN (1919) described some of the fresh-water *Molluscs* collected from campong Ngrawan and the banks of the Tjemoro river, but this is only a small part of the fauna occurring.

The list contains 11 species of which 4 (36½%) are still living. Though the number of fossils is too small to prove the age of the beds, it is evident that the fauna is different from **Trinil** (83% living species), only four species being identical. A Pliocene age is likely.

In considering the sequence of *Turritella* beds (Lower Pliocene), *Balanus* Limestone (Middle Pliocene,) an Upper Pliocene age for the fresh-water series is quite probable.

Only 4 species from **Bumiaju** are identical, two of these forms being restricted to **Sangiran** and **Bumiaju.**

*g. Diatom beds (intercalated in the black clay).*

The lowest *Diatom* bed increases in thickness from the SW (1½

| Fresh-water Molluscs from the Upper Pliocene of **Sangiran** | Miocene | Pliocene | Pleistocene | Recent | Marine Pliocene Sumberringin 1 | Terrestrial Pliocene Bumiaju | Terrestrial Pleistocene Trinil |
|---|---|---|---|---|---|---|---|
| Isidora sp. [1] . . . . . . . . . | — | + | ? | — | ? | | ? |
| Melania soloënsis Mart. [2] . . . . | — | (+) | — | — | | + | |
| „ tjemoroënsis Mart. [2] . . | — | (+) | — | — | | + | |
| „ bojolaliënsis Mart. [2] . . | — | (+) | — | — | | | |
| „ Fennemai Mart. [2] . . . | — | (+) | — | — | | | |
| „ granum v. d. Busch [2] . . | — | (+) | + | + | | | + |
| „ kritjianensis Mart. [2] . . | — | (+) | — | — | | | |
| „ testudinaria v. d. Busch[1] | — | + | + | + | | + | + |
| „ nov. sp. [1] . . . . . . . | — | + | — | — | + | | |
| Paludina javanica v.d. Busch [2] . | — | + | + | + | + | + | + |
| Corbicula exporrecta Mart. [1] . . | — | (+) | + | — | | | |
| Total species 11 . . . . . | — | 11 | 5 | 4 | 3 | 4 | 4 |
| Unio sp. and Melania sp. sp. indet. | | | | | | | |

M.) to the NE. (6 M.). The upper bed varies only from ½ to 1½ M. All beds are rather impure and contain a considerable quantity of argillaceous matter. The nature of the beds is easily recognizable by the low specific weight and the thinly bedded character, fine layers of *Diatoms* causing en excellent cleavability.

The *Diatoms* are only visible under the microscope when magnified about 100 times. They have not yet been determined as to genera and species.

*h. Tuffaceous layer (intercalated in the black clay).*

This layer is rather thin and not persistent. It proves the eruption of a neighbouring volcano and forms an excellent horizon to ascertain the dip in the generally badly stratified black clay.

*i. Yellow marine clay (intercalated in the black clay).*

The occurrence of a yellow marine clay of only ½ M. thickness is very remarkable, proving an invasion of the sea of short duration. The thickness of this bed is likewise exaggerated in the section and on the map.

The identical horizon occurs at a distance of 10 Km. northward

[1]) From a new collection by the author and determined by Dr. R. VON KOENIGSWALD.

[2]) determined by MARTIN (1919).

in **Baringinan,** where the thickness is likewise restricted and amounts to only 1½ M. The fauna is of marine habitus but very poor in species; on the other hand some of these are very abundant (indicated by *). The species determined from **Sangiran** are:

*Placuna placenta Linn.* *)
*Arca oblonga Phil.* *)
*Cominella sangiranensis Mart.*
*Melongena madjalengkensis Mart.* *)

As will be shown hereafter the impoverished fauna of the yellow clay of **Sangiran** and the yellow argillaceous sand of **Baringinan** show some resemblance to that of **Simo** 2,3 (60% living species), which probably is of Upper Pliocene age.

Pleistocene.

*k. Lower conglomerate-tuff series.*

The black clay is covered by an alternating series of conglomerates, sandstones, tuffs and clays together, mainly river deposits that formed the ultimate filling of the lake.

Though no unconformity is visible, as the succession of the beds is unbroken, it has been shown in the section, having been proved to exist in **Baringinan.**

The sequence of the beds is extremely variable.

It is to be noted that the coarseness of the material and the thickness of each individual layer seems to increase in a SE direction, corresponding to the direction from which the older volcanic boulder breccia originated.

The lower conglomerate-tuff series, together with the lower part of the upper conglomerate-boulder breccia-tuff series, from which it cannot be easily separated, form the main vertebrate horizon, in all probability corresponding to the **Trinil** beds.

Vertebrates especially abound in the vicinity of campong Bapang. The collection contains a lower jaw and several molars of *Stegodon*, molars of *Elephas*, horns of *deer*, and remains of *Crocodilus*, *turtles* and *fishes*.

Special mention must be made of a very remarkable discovery consisting of part of a molar wholly corresponding to that described by Martin (1887) from **Mt. Patihajam** and determined as *Mastodon sp.* but differing from the species occurring in **Bumiaju.**

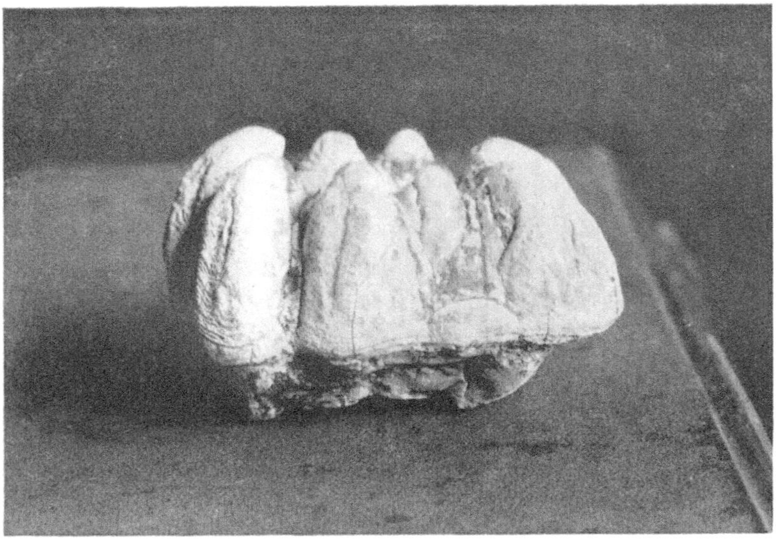

3. Part of a molar of Mastodon (?) from Sangiran
FRONT VIEW

4. Part of a molar of Mastodon (?) from Sangiran
TOP VIEW

*l. Upper conglomerate-boulder breccia-tuff series.*

The upper horizon consists of alternating conglomerates, boulder breccia, tuffs and sandstones and forms the tops and outward slopes of the circumferential ring of hills.

*Vertebrates* and *Unio sp.* are restricted to the lower horizons.

The higher horizons may be compared to the upper volcanic tuffs and conglomerates of **Mt. Patih-ajam** and of **Notopuro** (Kendeng mountains).

*Tectonic structure.*

The anticlinal dome of Sangiran in the circumferential zones shows simplicity in structure, the centre, however, being more composite. An E.—W. fault separates the northern part from the more complicated centre.

The dip of the beds on the NW. side is decidedly greater than that on the other side.

The impression was received that the whole structure was produced by a local push from the ESE. In the central part the pressure increased to such an extent that two culminations resulted, separated by a syncline. This explains the presence of almost horizontal beds of *Balanus* limestone and of boulder breccia in the centre of the dome.

Some steep hills of boulder breccia S. and W. of campong Sangiran, however, are to be excluded from this explanation. They are merely to be considered as erosial relics and do not really form part of the tectonic structure. The boulder breccia being more resistant than the underlying marine beds, big masses that had originally broken off from the circular outcrop of boulder breccia were brought to their present position by landslides.

In the western culmination the presence of an upthrust is to be deduced from the position of the beds on both sides of it. Moreover it is indicated by the presence of mud volcanoes.

Gas wells occur in both culminations, and in the western one especially in a circular ring around one of the mud volcanoes. The material brought to the surface by the mud volcanoes gave rise to a very interesting problem concerning the tectonic structure of the Tertiary mountain ranges in Java.

Besides the well-known fine mud derived from argillaceous Neogene beds, the products of the mud volcanoes contain blocks of some old andesite, quartz conglomerates and fragments and even a big block of *Nummulitic* (*Camerina*) limestone of Eocene age.

As the bituminous matter from which gas and oil wells originated is in Java generally supposed to occur in the Neogene beds, it is quite remarkable that older material should be brought to the surface. There are only two possible solutions to this problem. Either the supposition of a Neogene origin of the bituminous matter is untrue, or in this part of Java there exists a reverse position of the Tertiary beds, caused by an overthrust.

The author is inclined to accept the second solution as being more probable, as a marked unconformity between Upper Miocene and Eocene beds in the Djiwo mountains has been proved, which gives reason to the supposition of the deeper beds under the dome of Sangiran being even more disturbed than the surface layers.

### 8. **Kaliuter.**

(*With one map*).

*Situation*: River Kaliuter near campong Barong flows in a NE. direction and approximatively follows the boundary of Pliocene and Miocene beds. Campong Barong is a small village about 10 Km. N. of Kalioso.

*Stratigraphy.*

M i o c e n e.

*a. White Globigerina marls.*

The white *Globigerina* marls resemble those of S. Kendeng-mountains in the vicinity of Mt. Pandan. They are generally unstratified, but some few intercalations of sandy layers show a rather irregular dip of 30—50°. No further investigations were made as the position of the Pliocene and younger beds was considered to be of more interest.

*b. Globigerina limestone.*

The *Globigerina* marls pass upwards into *Globigerina* limestone of impure character. The dip was found to be generally 20° SE. They correspond to the Upper Miocene *Globigerina* limestone that will be afterwards described from Kendeng mountains.

*c. Sandy marls.*

Towards the NE. the *Globigerina* limestone disappears and is substituted by sandy marls owing to a greater content of impurities.

*d. Gray banked sandy marls (Transition beds).*

The higher layers are to be considered as *transition* beds between the Upper Miocene and the Lower Pliocene. They still contain *Globigerines* to such an extent that a sort of stratification is present due to a varying content of lime in alternating beds. The age of these beds of course was not proved in **Kaliuter.** The same succession, however, being present in the Kendeng mountains near Mt. Pandan, a corresponding age was assumed by the author by way of analogy.

Lower Pliocene.

*e. Blueish gray argillaceous beds.*

The argillaceous beds are generally badly stratified and of a dark blueish gray colour. *Globigerines* are almost absent and small sized. The beds much resemble the dark Lower Pliocene argillaceous layers **N. of Kendeng mountains** and of **Sangiran.** The total thickness of this series is estimated at about 150 M.

*f. Molluscbearing conglomerate.*

A superincumbent bed of conglomerate contains pebbles of andisitic origin together with a fauna of *Molluscs.*

| *Molluscs* from **Kaliuter** collected by Dr. K. G. SCHMIDT and determined by Dr. R. VON KOENIGSWALD | Miocene | Pliocene | Recent | Sangiran Lower Pliocene | Bareng Middle Pliocene | Sonde Middle Pliocene |
|---|---|---|---|---|---|---|
| *Gasteropods* | | | | | | |
| Surcula nodifera Lamk. . . . . | + | + | + | | + | |
| Drillia neglecta Mart. . . . . . | + | — | — | + | | |
| Marginella quinqueplicata Lamk. | + | + | + | + | + | + |
| Metula hindsii H. e. A. Ad. . . . | + | + | + | | | |
| Phos acuminatus Mart. . . . . | + | — | — | | | |
| Cassis herklotzi Mart. . . . . . | + | + | — | | | |
| Ranella subgranosa Beck.. . . . | + | + | + | + | + | + |
| „ pulchra Gray. . . . . . | + | + | + | + | + | |
| Solarium perspectivum Linn. . . | + | + | + | + | + | + |
| Natica globosa Chemn.. . . . . | + | + | + | + | + | |
| *Lamellibranchiates* | | | | | | |
| Limopsis venusta Mart. . . . . | ? | + | — | + | + | |
| Total species 11 . . . . . . | 10+1 | 9 | 7 | 7 | 7 | 3 |

The list contains 11 species, 7 of which (or 63%) are still living. A considerable number of *Molluscs* remained undetermined yet, several of which probably are new species. This necessitates some caution as to conclusions about the age of this fauna from the results of the published list. Whilst the high percentage of living species would point to a rather young age, on the other hand all eleven determined species are known to occur in the Miocene (one is uncertain), against only 9 species that are also known from Pliocene beds. One of the apparent Miocene forms, however, is described from the Lower Pliocene of **Sangiran.**

The succession of the beds in **Kaliuter** greatly resembles that in **Kendeng mountains.** Stratigraphically the blueish gray argillaceous beds correspond to the Lower Pliocene argillaceous beds in that region. Therefore the author considers the fauna of **Kaliuter** as representing a fossiliferous horizon of the Lower Pliocene. Though not quite corresponding to the *Turritella* bed of **Sangiran** as is shown by the small number of corresponding species (considering the undetermined forms), both horizons cannot much differ in age as they are underlying the *Balanus* limestone.

Though the same number of corresponding species is found in **Bareng** and **Sangiran,** one has to consider the fact that the general character of the fauna certainly points to a greater age than the Middle Pliocene. There is very small resemblance to **Sonde** where only 3 similar species occur.

Altogether, the *Mollusc* fauna of **Kaliuter** is considered by the author as Lower Pliocene, but this age is more concluded from the *lithologic character* of the beds and not from the fauna, which is insufficient for a convincing conclusion.

*f. Greenish gray argillaceous sandstone.*

The beds overlying the fossiliferous horizon of **Kaliuter** consist of argillaceous sandstones that have a thickness of about 100 M.

Middle pliocene.

*g. Balanus limestone.*

The described Lower Pliocene argillaceous and sandy beds are unconformably covered by *Balanus* limestone of a habitus quite corresponding to that of **Sangiran.**

The outcrop of *Balanus* limestone does not run parallel to the bedding of the Lower Pliocene beds. It appears that in the S. part of the map the thickness of the Lower Pliocene beds exposed amounts to 250 M., whereas near Barong it is reduced to only 50 M. This has been caused by the unconformable position of the *Balanus* limestone. Moreover this limestone bears all the features of having been deposited during a transgression and at the base it often contains pebbles of andesite showing its conglomeratic character.

The thickness is greater than in **Sangiran** and varies from 5 to 15 M. As it obviously represents the same bed as in **Sangiran** the age is considered as Middle Pliocene.

Upperpliocene?

*h. Fresh-water sandstone and tuff.*

The superincumbent beds are seldom well exposed. The visible part of it consists of fluviatile tuffs and sandstones of volcanic origin. Probably they correspond with the lake deposits of **Sangiran.**

Pleistocene.

*i. Conglomerates and boulder breccia.*

The next well-developed beds consist of conglomerates and volcanic boulder breccia unconformably overlying the Miocene and Pliocene beds.

SE. of the river Kaliuter, the boulder breccia covers the *Balanus* limestone and the fluviatile bedded tuffs and sandstones.

On the NW. bank however the boulderbreccia lies on top of Miocene white *Globigerina* marls and *Globigerina* limestone or covers the Lower Pliocene beds. A difference in level on both banks shows that a slight dip in the boulder breccia is present, probably caused by tilting movements.

The age probably is the same as that of the highest tuff and boulder breccia horizons in **Sangiran.** No *Vertebrates* occur, so it seems that the typical *Vertebrates* of the **Trinil** horizon are absent.

Tectonics.

The main result of the investigations is the discovery of two unconformities: one between the (Middle) Pliocene *Balanus* limestone and the Lower Pliocene argillaceous beds and the second between the Pleistocene conglomerates and boulder breccia and the underlying beds of Miocene and Pliocene age.

## 9 **Baringinan.**

(*With one map*).

*Situation.* The region is situated 10 Km. NNE. of Sangiran.

*Stratigraphy.*

M i o c e n e.

*a. Alternating tuffs and marls.*

The deepest beds exposed, consist of well-bedded marls and tuffs or sandstones, resembling in habitus the deeper horizons of the Miocene in Kendeng mountains near Mt. Pandan.

They evidently occur in the axial zone of two anticlines. The southernmost anticline shows dips, that indicate the presence of an overthrust fault on the northern slope. The difference in dip in adjacent sections, however, proves the presence of transverse faults, one of which moreover was proved to exist by the difference of beds on both sides of it.

*b. White Globigerina marls.*

The superincumbent layers resemble the white *Globigerina* marls of Kendeng mountains and only seldom show stratification. The synclinal position of the *Globigerina* marl N. of Randuguntіng therefore is not te be deduced from the scarce dips, but can be concluded from the relative position to the underlying tuffs and marls.

*c. Globigerina limestone.*

The next horizon consists of *Globigerina* limestone. There is certainly more than one layer present but as it was difficult to ascertain the habitus of the interjacent beds, the latter was not indicated on the map. An irregularity seems to be present between Teguhan and Tempel which indicates the presence of a third anticlinal region.

*d. Lower marine argillaceous sandstone* (*Transition beds*).

The habitus of this bed somewhat resembles that of the transition beds between Miocene and Pliocene in Kaliuter and the Kendeng mountains near Mt. Pandan.

L o w e r P l i o c e n e.

*e. Lower marine argillaceous beds*

Argillaceous beds of dark colour greatly resembling the Lower Pliocene argillaceous beds, occur on the eastern part of the map. Towards the W., however, they seem to wedge out or to pass into argillaceous sandstones somewhat resembling the underlying bed with this difference however, that *Molluscs* occur. On the other hand there are some signs of an unconformity being present and it is possible that this caused the apparent thinning out of the argillaceous strata.

Middle Pliocene.

*f. Balanus limestone.*

The *Balanus* limestone closely resembles that of **Sangiran** and **Kaliuter** and is generally well developed. It likewise contains pebbles of andesite and has a transgressive character. The thickness is variable. The occurrence of *Balanus* limestone directly covering the *transition* bed of marine argillaceous sandstone near campong Djuring indicates the probability of an unconformity.

*g. Gray marine argillaceous beds.*

The overlying gray argillaceous beds still are of marine origin. In **Kaliuter** and **Sangiran** this horizon is lacking.

Upper Pliocene.

*h. Lower argillaceous fresh-water beds.*

There is no sharp transition from the marine argillaceous beds to the overlying dark clay that appears to be a lake deposit.

The Molluscs at the base are restricted to fresh-water *Corbicula exporrecta*, occurring in great number.

The higher horizons contain a number of other fresh-water fossils.

The succession of the beds and the corresponding fauna show that this horizon is identic to the *Corbicula* bed of **Sangiran.**

*h. Yellow argillaceous sand with marine Molluscs.*

The yellow argillaceous sand corresponds to the yellow marine clay observed at **Sangiran,** and likewise represents the relic of a marine invasion in the Upper Pliocene lake of Northern Suracarta The thickness of the bed increased however to 1,5 M.,which probably is due to the coarser character of the deposit.

The resemblance to **Sangiran** is emphasized by the *Mollusc* fauna, several species not only being identic but likewise occurring abundantly.

It was however still more important to notice the resemblance of this fauna to that collected by Dr. J. M. W. NASH from **Simo** 2 and 3 in N. Kendeng mountains N. of Djombang, that will be described later on. There is also some resemblance to the fauna of the *Turritella* horizon of **Baribis.**

Next to marine species of *Molluscs* and *Rotalia sp.*, the argillaceous sand with marine *Molluscs* of **Baringinan** contains the following fresh-water species:

*Paludina javanica v. d. Busch, Melania testudinaria v. d. Busch, Lymnea sp., Isidora sp., Planorbis sp., Segmentina sp.* and *Hydrobia sp.*

It is evident, however, that the fresh-water *Molluscs* were derived from the underlying beds in which they are very frequent.

| *Molluscs* of the impoverished marine fauna of the yellow argillaceous sand of **Baringinan** collected by Dr. K. G. Schmidt and determined by Dr. R. von Koenigswald | Argillaceous sand. Baringinan | Yellow clay Sangiran | Simo 2, 3 | Baribis Turritella horizon | Recent |
|---|---|---|---|---|---|
| *Gasteropods*: | | | | | |
| Marginella quinqueplicata Lamk. | + | | + | | + |
| Cominella sangiranensis Mart.. . | + | + | + | | |
| Melongena madjalenkensis Mart. | +[1] | +[1] | +[1] | | |
| Siphonalia paradoxica Jenk. . . | + | | | | |
| Purpura carinifera Lamk. . . . | + | | | | + |
| Turritella terebra Lamk.. . . . | + | | + | +[2] | + |
| *Lamellibranchiates*: | | | | | |
| Placuna placenta Linn. . . . . | +[1] | +[1] | +[1] | +[1] | + |
| Chlamys kebolintangensis Mart.. | + | | | | |
| Anomya sp. . . . . . . . . . | + | | | | |
| Arca oblonga Phil. . . . . . . | + | +[1] | +[1] | +[1] | + |
| Arca inaequivalvis Brug. . . . | + | | + | | + |
| Arca nov. sp. . . . . . . . . | + | | | | |
| Meretrix cf. meretrix Lamk. . . | + | | | | + |
| Corbula socialis Mart. . . . . . | + | | + | + | |
| Venus bataviana Mart.. . . . . | + | | | | |

The number of determined species amounts to 13, 54% of which are still living. Though a Pliocene age can be deduced, the number is too small to ascertain the exact horizon. The succession of the beds, however, shows that an Upper Pliocene age is more likely, as the horizon certainly is younger than **Sonde**.

*i. Upper argillaceous fresh-water beds.*

The overlying beds still contain fresh-water Molluscs and appear

[1]) Abundantly occurring.

[2]) The Turritella, from Baribis with a marked solitary keel, is perhaps only a variety of Turritella terebra.

to be lake deposits. The fauna is not so rich however as in **Sangiran.** Remains of *Vertebrates* are present.

P l e i s t o c e n e.

*j. Sandstones, tuffs, conglomerates, black clay and volcanic boulder breccia.*

The lake deposits and the *Balanus* limestone are unconformably overlain by conglomerates, tuffs and boulder breccia.

The unconformity is quite evident as in the northernmost exposures N. of **Baringinan** patches of conglomerates directly overlie the *Balanus* limestone or even Lower Pliocene marine argillaceous beds, whereas on the southern border much younger horizons of lake deposits are covered. The conglomerates and volcanic sediments represent the highest horizon of the beds in **Sangiran.** Vertebrate remains do occur, but are rather scarce.

10. **Soloriver between Gesi and Ngawi. (Trinil).**

(*With one map and one section*).

*Situation.* The map represents the course of the Soloriver between Gesi, N. of Sragen and Ngawi. The river almost follows the boundary between the Tertiary beds and the Quaternary bone-bearing horizons. It passes repeatedly from one formation into the other and thereby produces some very good exposures on its banks.

*Stratigraphy.*

M i o c e n e.

*a. Alternating tuffs and marls, volcanic breccia and limestone.*

The deepest beds occurring in this region are alternating tuffs and marls with a few intercalated beds of volcanic boulder breccia and of limestone. This group of layers corresponds to similar beds in the Kendeng mountains in the vicinity of Mt. Pandan. The limestone contains small *Foraminifera* belonging to species of *Lepidocyclina* and *Miogypsina*. This proves the beds to represent the division indicated by VAN DER VLERK and UMBGROVE as Tertiary f.

*b. White Globigerina marls.*

The next horizon is a complex of white *Globigerina* marls, that resembles the corresponding beds in the Mt. Pandan-region, where an Upper Miocene age is evident. The fauna is restricted to *Globigerines* and *Diatoms* and the stratification generally is rather bad.

*c. Coral limestone.*

Limestone beds occur intercalated in the white *Globigerina* marls W. of Mantingan. But the *Coral* limestone that covers the white *Globigerina* marls between Ngawi and Watugudel wholly corresponds to the Upper Miocene *Coral* limestone of Kendeng mountains in the vicinity of Mt. Pandan.

*d. Transition marls.*

In the neighbourhood of Sonde and Bener, the uppermost layer of the Miocene consists of gray sandy marls, resembling the transition beds that already have been described from **Kaliuter** and **Baringinan.** Though they still contain *Globigerina*, they differ from the white *Globigerina* marls by a greater content of sandy impurities.

Lower Pliocene.

*e. Lower argillaceous beds.*

In the extreme Western part of the region mapped, S. of Gesi argillaceous beds appear, that underlie Pliocene *Coral* limestone and cover white *Globigerina* marls. They are very similar to the Lower Pliocene argillaceous beds known from **Sangiran, Kaliuter** and **Baringinan.** The Middle Pliocene beds being known to overlie unconformably all older formations, the discontinuity in the outcrop of several horizons is not astonishing at all. The transgression is accompanied by a slight unconformity and it is only at the extreme Western part of the map, not far distant from **Baringinan,** that the Lower Pliocene beds are exposed, whereas everywhere else they were either never deposited or are covered.

Middle Pliocene.

*f. Conglomeratic beds and Coral limestone.*

In campong Bener, beds of bedded limestone where discovered resembling the basal bed of the Pliocene in the region E. of Ngawi and near Mt. Pandan.

It contains pebbles of Miocene *Globigerina* marls and is to be considered as a basal conglomerate indicating a transgression of the Middle Pliocene.

Near Bener this conglomeratic limestone is covered by beds of limestone alternating with marls containing *Molluscs*. The limestone and the marls probably represent a somewhat changed facies of the **Sonde** beds.

The *Coral* limestones W. of Bandjarbangi are the equivalent of *Coral* reefs that occur as intercalations in the **Sonde** beds. The Upper Miocene and the Pliocene *Coral* limestones are easily confused. The habitus is similar and the fauna has not been sufficiently studied yet to distinguish both horizons by means of fossils. The author in a map published in 1929, made an erroneous statement in assuming all *Coral* limestones in this region to be of Pliocene age. The author was unacquainted then with the fact, that the marine Pliocene, represented by the characteristic **Sonde** beds, occurs there in a very reduced thickness not exceeding 50 M. The Miocene *Globigerina* marls likewise were erroneously represented as Pliocene.

The Pliocene being transgressive, its base is characterized by a conglomerate, which in some cases may serve as a means to identify the age of the *Coral* limestone. The limestone near Bener was mapped as Pliocene because the basal bed contains pebbles of Miocene *Globigerina* marls.

*g. Argillaceous sand (Sonde beds).*

The exposures of argillaceous sand or sandstone, containing Molluscs are restricted to the neighbourhood of **Sonde, Padasmalang** and **Pengkol,** and to the extreme Western part of the map S. of Gesi.

In **Sonde** a rich fauna of *Molluscs* is present. MARTIN (1919) determined 150 different species, 53% of which were found to be still living, proving a (Middle) Pliocene age. According to FELIX (1911) of 11 *Corals* occurring in **Sonde,** 72% belong to living species, whilst of 39 *Corals* found in **Pengkol** and 14 *Corals* of **Padasmalang,** 64% belong to recent species. This slightly higher percentage of living species of *Corals* as compared to the *Molluscs* needs no further explanation. For different groups of fossils another percentage of living species is to be expected in every case.

A number of Molluscs occurring in **Padasmalang** and **Pengkol** was described by MARTIN—ICKE (1911) and was formerly kept separate from the list of **Sonde**. The field work proving those occurrences to represent quite the same horizon, the author judges it no longer necessary to continue this separation. Moreover the author collected the following species from similar beds in the Mt. Pandan-region near **Sukun** and **Soko,** that were not included in the above mentioned lists:

*Cominella sangiranensis Mart.*
*Ostrea lingua. Sow.*
*Placuna placenta Linn.*
*Corbula socialis Mart.*

This would increase the number of species from the **Sonde** beds to 179 of which 99 or 55,3% are still living.

The total thickness of the beds does not exeed 50 M.

The apparent disappearance of the **Sonde** beds E. of Trinil has to be ascribed to an unconformity existing between the Pliocene and overlying Pleistocene beds. Likewise near Mantingan the Miocene tuffs and marls are directly covered by Pleistocene conglomerates, volcanic breccia and sandstones, and the Pliocene *Molluscs*-bearing beds are not exposed. But in a borehole just South of Mantingan they were met with at a depth of 115 M., showing that their apparent absence is due to an unconformable position of the overlying Pleistocene beds.

P l e i s t o c e n e.

*h. Sand and conglomerate containing older material.*

South of campong Bener the Pliocene limestone directly is covered by conglomerates exclusively consisting of limestone pebbles. They contain fossil bones and probably correspond with bone-bearing (*Stegodon*) conglomerates at Bandjarbangi and a limy sand that covers the **Sonde** beds in **Pengkol.** The latter contains small pebbles and detritus of Miocene beds. These beds seem to indicate a period of denudation preceding eruptions of Mt. Lawu.

*i. Volcanic boulder breccia.*

In the region between **Ngawi** and **Sonde** volcanic boulder breccia occurs. It generally underlies the main vertebrate horizon represented by the rich bone-bearing bed of **Trinil.** This breccia corresponds to a similar horizon E. of Ngawi, which is very persistent. Several beds of volcanic breccia occur as intercalations in the bone-bearing beds in the Western part of the map. They show an increased volcanic activity of Mt. Lawu during the Lower Pleistocene.

*j. Conglomeratic sandstones, conglomerates, sandstones, tuffs, and black clay (Trinilbeds).*

This very variable succession of beds is of the greatest interest as it represents the main *Vertebrate* horizon.

The very rich bone-bed in **Trinil** has been differently interpreted by various authors. To the opinion of the author it is to be considered as a river deposited conglomeratic tuff sandstone. A list of the fauna occurring in this horizon has already been given in Chapter IV.

The bones seldom occur as abundantly as in **Trinil.** As a result of investigations made in 1926 and 1927 the author drew the attention of the 4th Pacific Science Congress (1929) to the rich occurrences of *Vertebrate* bones in **Watualang,** in **Bandjarbangi** and in **Tawang-Kedungkendang,** that might compete with the bone-bed at **Trinil.**

The origin of the bone-beds sometimes has been ascribed to the destruction of the existing fauna by volcanic eruptions. However a thorough investigation gave the following results:

In many cases the bones were found in cross-stratified sandstones containing rounded pebbles, that are to be considered as true river deposits. It is to be supposed that the animals died through natural causes.

The decaying carcasses were swept by flooded rivers to some wirlpoolbend where they sank or to sandy banks where they finally decomposed. Generally the bones are found apart and in several cases were already broken and weathered before they were buried in the sand. At times, however, parts of vertebral columns were found to occur in the original position, which shows that transport in some cases is totally excluded.

Crocodiles living in deep waterholes often caused an accumulation of bones.

In other cases the bone-bearing bed consists of black clay, containing fresh-water *Molluscs* and remains of *fishes, crocodiles* and *turtles*. This black clay has been formed in stagnant pools and marshes or even in big lakes. Sometimes nearly complete skeletons of larger *Vertebrates* occur, owing to the fact that marshes often form the dwelling place of big animals. Whilst the animal was dying the carcass may have been swallowed by the mud.

Loose bones, and even whole skeletons may have been covered by volcanic ashes, but this does not lead to accumulations of bones. On the contrary the frequency of bones in the **Trinil** beds excludes a strong volcanic activity. Erosion certainly was predominant.

This is shown by the fact that the **Trinil** beds, though for the greater part composed of water-transported volcanic matter, often contain detritus of Miocene beds in which sometimes even the *Globigerines* are preserved. The *Globigerines* deceived some authors who stated corresponding beds in the vicinity of Mt. Pandan to be partly of marine origin.

The detrital products of the Miocene often are responsable for the lime content of the Trinil beds to which the solidification of the sandstones and conglomerates is due.

The observation that the **Trinil** beds were partly derived from older beds arouses the question of the origin of the bones. In admitting the preponderant influence of erosion it would not be inconceivable that the bones likewise were derived from destroyed older beds. Indeed, the fact that many bones are broken and partly rounded by rivertransport would confirm this supposition. But on the other hand it is also to be stated, that other bones do not show traces of wear. Fossil bones and especially fossil ivory are too brittle to suffer any transport. This certainly excludes an older origin for the *Stegodon* tusks and for the skulls of horned beasts that still possess the protuberation of the horns. Several discovered lower jaws and molars of *Stegodon* likewise are too well preserved to accept their being much transported. Finally, the bigger bones and skulls often occur in mediumgrained sand containing only small pebbles, whilst big boulders are absent. In such cases a transport of the bones in a fossil state is hardly conceivable, as the specific weight increases by fossilization.

*Pithecanthropus erectus* was discovered in **Trinil** in a bone-bearing bed exposed along a bend of the Soloriver. This bed has been excavated on the Eastern bank down to the lowest level of the river by DUBOIS and the SELENKA expedition. The latter tried to continue the excavations on the western bank, but it seems that the bone-bearing bed wedged out in this direction. To the East no sufficient attempts have been made to rediscover the bed.

Tectonics

The evidence for a Pleistocene age of the **Trinil** beds will be discussed in the final Chapter. The main result of the investigations in this region is the conclusion that the **Trinil** beds and the volcanic boulder breccia unconformably cover the Miocene and Middle Pliocene beds. The volcanic boulder breccia and the conglomerates often directly cover the Miocene beds, which explains the apparent absence of the Pliocene **Sonde** beds in the neighbourhood of Ngawi and Mantingan.

The Trinil beds generally dip 6—10° S. which shows the influence of tilting or folding movements. Moreover several transverse faults are present, showing that the beds did not remain undisturbed.

A second unconformity separates the Middle Pliocene from the Miocene and Lower Pliocene beds, and accounts for the very small thickness of the Middle Pliocene beds, that do not exceed 50 M.

## 11. Southern border of Kendeng mountains between Ngawi and Redjuno.

(*with one map and one section*).

*Situation*: The region described is the low hilly country forming the southern border of Kendeng mountains east of the transverse valley N. of Ngawi, were the Solo river breaks through the hilly range. To obtain an insight in the geological conditions, more than 10 sections have been thoroughly investigated between Ngawi and Redjuno.

*Stratigraphy*.

Uppermiocene.

*a. Globigerina marls.*

The deepest levels established in Kendeng mountains do not figure in the map. Only the younger white *Globigerina* marls with only a few alternating layers of volcanic tuffs or sandstones are represented.

*b. Globigerina limestone.*

The overlying beds consist of *Globigerina* limestone, only differing from the marls by a smaller content of argillaceous or tuffoid matter. The greater content of lime generally makes them more solid, but on the other hand more porous, as they are almost exclusively built up by the very porous shells of *Globigerines*. This horizon is lacking E. of the transverse valley of Dero-river, but probably this is simply a matter of facies. It is quite possible that the impurity of the identical bed makes it undistinguishable from the underlying *Globigerina* marls.

*c. Coral limestone.*

The higher beds consist of *Coral* limestone. As is to be expected the thickness of the *Coral* reef is very variable and the horizon is even absent in the vicinity of the transverse valley of Kali Pang.

A few *Molluscs* appeared to be determinable and to belong to the following species:

| *Molluscs* from *Coral* limestone, determined by the author | Miocene | Pliocene | Recent |
|---|---|---|---|
| Cassis pila Reeve . . . . . . | ? | + | + |
| Voluta scapha Gmel var. . . | + | + | — |
| Ostrea lingua Sow. . . . . . | + | + | — |
| Tridacna gigas Linn. . . . . | + | + | + |
| Tridacna rudis Reeve . . . . | + | + | + |

The number of species, of course is too small to determine the age of the beds. *Tridacna gigas* occurs rather abundantly in the region N. of Redjuno. The upper part of the limestone generally has a conglomeratic character. As it is exclusively composed of limestone pebbles and no other matter is mixed with it, this conglomerate probably was formed by the surf contemporaneous to the formation of the *Coral* reefs.

*d. Transition beds.*

The Upper Miocene transition beds, that already have been described from the areas of Kaliuter, Baringinan and Gesi-Ngawi, reappear over a short distance between Ngawi and Kionton. They consist of sandy matter and contain some Globigerines.

Middle Pliocene.

*e. Marine argillaceous sand (Sonde beds).*

In the eastern part of the map, starting from the vicinity of Dero the Miocene beds are covered by darkgray marine argillaceous sandy beds containing *Molluscs*. The shells however are generally badly preserved, therefore the palaeontologic proofs for a Middle Pliocene age are lacking.

On the other hand the *lithologic character* of the beds closely resembles the **Sonde** beds. As the interruption in the outcrop of the **Sonde** beds W. of Ngawi was due to an unconformable covering by Pleistocene beds, it is evident that somewhere E. of this place a reapparition of this horizon was to be expected. The thickness of the beds is likewise very restricted.

Upper Pliocene.

*f. Lower breccia and lower tufflayers* (*Butak beds*).

In the vicinity of Redjuno E. of Kali Pang the marine argillaceous beds are covered by volcanic breccias alternating with tufflayers. The lowest breccia horizon contains marine *Molluscs*. As a volcanic breccia during its deposition easily may have derived the shells from underlying beds — in the same way as has been described in Sangiran — there is not sufficient evidence of the breccia being a marine deposit. The bulk of the lower breccia and lower tufflayers is certainly of terrestrial origin and was formed by volcanic activity of Mt. Wilis and Mt. Pandan. Though of course the thickness decreases the farther the distance is from those volcanoes, there are signs of the overlying beds unconformably resting upon Pliocene and Miocene beds; the apparent lacking of the lower breccias and tuffs W. of Kali Pang might be ascribed to this unconformity. On the other hand it is to be admitted that there is no certainty of the higher horizons of this group of beds being still of Tertiary age.

Pleistocene.

*g. Middle tufflayers and upper breccia.*

The upper breccia horizon, persistant in the whole area mapped, is accompanied by an underlying bed of tuff between Kionton and Pantjuran. It appears that these layers already belong to the group of beds, that unconformably cover the Miocene and Pliocene and therefore a younger age than the lower part of the **Butak** beds is likely. In the neighbourhood of Ngawi they directly overlie the Miocene transition beds, whereas E. of Dero they rest upon the Middle Pliocene **Sonde** beds. Near Redjuno they are to be supposed as covering the **Butak** beds, from which they cannot be well separated however, as the habitus of both groups of layers is quite similar.

*h. Conglomerates and sands (Trinilbeds).*

The conglomerates and sands, that overlie the upper breccia horizon, contain *Vertebrates* and in many features resemble the **Trinil** beds W. of Ngawi. *Vertebrates* abound N. of Redjuno and occur in the low hills between Dero and Boto and on the banks of the Solo and Madiun rivers near Ngawi. In some of the conglomerates, pebbles of Miocene limestone occur, in which *Globigerines* and small *Foraminifera* still could be traced.

In the low almost flat country between Boto and Dungpatjul the **Trinil** beds are covered by a black clay originating from the weathering of these beds, and therefore do not show good exposures.

*i. Upper tuff layers.*

South of Redjuno and near Ngawi the layers covering the **Trinil** beds are exposed. They consist of tuff beds that are to be considered as representing the **Notopuro** beds in the vicinity of Mt. Pandan.

## 12. Kendeng Mountains in the vicinity of Mt. Pandan.

*(With one map and 7 sections).*

*Situation:*

The name Kendeng mountains is given to a low tertiary-folded hillrange in the axis of the eastern broad part of Java. It extends from the region S. of Semarang to the vicinity of Surabaja, where the height decreases to mere low hills. N. of Ngawi the Solo river breaks through it.

The investigated region lies W., N. and E. of Mt. Pandan, the top of which forms the junction of the boundaries of the three residencies: Madiun, Kediri and Bodjonegoro (formerly Rembang).

The main purpose of the investigations being the determination of the age of the Vertebrate series, the Miocene beds, forming the central zone of the mountain range, have been mapped fragmentarily only.

The watershed generally is near the southern border of the range. Some of the southward flowing rivers however succeeded in breaking through the southern border consisting of more solid volcanic and Vertebrate series and start from the softer Miocene beds of the central zone.

The orography in the Tertiary beds is dominated by E-W. ridges caused by more solid beds, the strike of which is fixed by the tectonic structure. Sometimes the abrupt ending of the ridges is due to the presence of transverse faults.

Mt. Pandan does not really form part of Kendeng mountains as it lies at some distance S. of it. It certainly is the highest peak in the whole region (900 M.), whilst the ridges of Kendeng mountains generally do not surpass 350—450 M.

North of Mt. Pandan some striking steep peaks or „kopjes" — formed by volcanic dykes — occur.

*Stratigraphy.*

M i o c e n e.

*a. Breccia, tuffs, sandstones, marls.*

The oldest beds, exposed in the axis of some anticlines of Kendeng mountains, consist of well-stratified, greenish to yellowish marls, alternating with tuffs, sandstones and breccias. Generally no fossils are present with the exception of *Globigerines* and *Diatoms*, but in a limestone from G. Gudeg, interbedded in this series, *Nephrolepidina sumatrensis Brady* was found. The limestone occurring on the slope of G. Diatasangin is probably of the same age. The mentioned fossil likewise occurs in a limebearing breccia south of Tempuran. It defines the age of the beds as the upper part of Tertiary *e* or the lower part of Tertiary *f*.

The sandstones show cross-stratification and, owing to basic volcanic grains, a brown weathering colour. The coarse tuffs often pass into volcanic breccias. A thick bed of volcanic boulder breccia occurs NE. of Sukun and N. of Putuk. With increasing distance from the region N. of Mt. Pandan, the breccias and sandstones become finer-grained.

This would give some hold to the supposition that volcanism during the Lower Miocene was active in the vicinity of Mt. Pandan. But the author thinks it necessary to point out that Mt. Pandan in its present form undoubtedly is a slightly eroded Quaternary volcano, the products of which for a great deal are even younger than the Vertebrate series.

Volcanic boulderbreccias occurring in the Miocene have already been described from the region between Gesi and Ngawi, where evidently no volcanic centres occur.

For these reasons the Miocene volcanism is not to be confounded with the later Upper Pliocene and Quaternary Mt. Pandan volcano. The main region of volcanic activity during the Miocene is certainly to be sought farther south in the mountain ranges of southern Java.

The Miocene tuffs of Kendeng mountains are to be considered as volcanic air- and water-transported products of southern origin deposited in a northern neritic geosynclinal sea.

This geosyncline was situated between a continent or a group of islands in the North now covered by the Java Sea and the geoanticlinal and volcanic region of Southern Java.

As has been pointed out by RUTTEN (1925) the lithological character of the Tertiary beds is different in the North and in the South of Java. The beds exposed in the anticlines N. of the line Tjepu-Bodjonegoro mainly consist of detrital sediments of Northern origin, whereas a tuffoid habitus prevails in Kendeng mountains.

In the region N. of Soko alternating series of badly stratified gray *Globigerina* marls occur between well-bedded mainly volcanic sediments pointing to the possibility that the breccia-tuff-sandstone-marl beds are partly contemporaneous to the Upper Miocene gray *Globigerina* marls.

In the SE. part of the investigated region N. of Sendenggogor *Diatom* beds occur intercalated between the tuffs and marls. They correspond with the *Diatom* beds of Kendeng mountains N. of Djombang. As in the latter region the *Diatom* beds alternate with *Globigerina* marls, this strenghtens the above supposition.

*b. Gray Globigerina marls.*

Especially in the Northern anticlines of Kendeng mountains the beds occurring in the axis consist of sometimes strongly compressed and steeply folded gray *Globigerina* marls. These gray marls are seldom developed in the southern anticlines.

The stratification is generally bad, but sometimes the different habitus of intercalated beds makes it rather easy to ascertain the dip. Intercalated in the gray marls there occur beds of a whiter colour, possessing a greater content of lime, caused by a more abundant occurrence of *Globigerines*. In the gray marls the *Glo-*

*bigerines* are generally scarce, but only in a very few places these marls are entirely barren of *Foraminifera*.

In weathering the gray marl shows a nodular disintegration, the nodules falling to pieces which resemble the scales of an onion.

Sometimes the stratification of the marls can be observed from the arrangement of the weathered nodules, and at other times the bedding seems to be exceptionaly good, but the author is not certain that this is not a result of strong pressure, and more corresponds to cleavage.

The whole thickness of the gray marls is estimated at several hundred meters. But it is very difficult to ascertain the true figures as the beds often are strongly compressed and faults may occur which have been overlooked in the field.

It is a remarkable fact that generally the gray marls fail to occur in the southern anticlines. This might be explained by a difference in facies. The gray marls perhaps partly correspond with the well-stratified marls alternating with tuffs and sandstones in the southern anticlines. But on the other hand it is probable that they partly correspond with some of the white marls known in the southern anticlines.

Just North of G. Puntju it is evident that white and gray marls alternate in such a way that it is difficult to say to which horizon this series of beds is to be ascribed.

Nevertheless it is evident that there is a considerable difference between the sequence of the bed in the Northern and the Southern anticlines. Though this is to be ascribed to a difference in facies, it is a fact that the change is rather abrupt and wants more explanation.

NE. of Mt. Pandan on the banks of R. Patjal between Sukun and Gondang there occur gray marls of a somewhat softer nature than elsewhere is the case. They somewhat resemble the Lower Pliocene argillaceous beds.

They seem to contain *Ostrea lingua* near Gandong and Sambungredja. But as this fossil was found in some loose specimens only, on the spots indicated on the map, and the author discovered that in some other cases the shells are used by the natives for burning lime, there is not sufficient evidence to regard this observation as a proven fact.

*c. White Globigerina Marls.*

White *Globigerina* marls occur both in Northern and in Southern anticlines. The transition of the gray into white marls in the northern anticlines is gradual and is caused especially by a greater amount of *Globigerines,* whereby the limecontent is increased. At first white and gray marls alternate and ultimately the white marls predominate.

The increase of the *Globigerines* may be partly due to a more vital plankton-life or to currents supplying the shells. On the other hand the amount of volcanic material being less, this caused a greater percentage of the marine deposits to consist of *Globigerines.*

This corresponds with the results of investigations in the southern anticlines, where the underlying formation shows a more or less volcanic character, whereas in the white marls, tuff is restricted to a few intercalating layers. This is further wholly confirmed by the facts found in G. Sewu in Southern Java where the author discovered the Upper Miocene sediments to consist mainly of *Globigerina* limestones, *Lithothammium-* and *Coral*-reefs, with only a few intercalating tuff beds.

Volcanic action certainly decreased during the Upper Miocene and — with only a few exceptions — even ceased to exist.

*d. Globigerina limestone.*

The decrease of volcanic matter even went so far, that the chief material deposited consisted of *Globigerines* forming *Globigerina limestone.* These limestones are not quite free from volcanic matter, however, and still show a few grains of augite.

Well-banked limestones of this type occur in the northern anticlines at the top of the white marls, for instance S. of Tretes and on the slopes of the syncline of Maor. They are still covered by at least 100 M. greyish white transition beds. Near Bareng the banked *Globigerina* limestone is lacking. On the other hand the white *Globigerina* marl is covered there by an impure *Globigerina* limestone, containing rather more volcanic matter, that, however, is to be considered as belonging to the transition beds.

In the Southern anticlines the *Globigerina* limestone is well exposed near G. Puntju where it forms the main body of an E-W. ridge.

*f. Coral limestone.*

In the Southern anticlines the white *Globigerina* marls are generally covered by a *Coral* limestone. It is often quite pure and sugar-white, the impurities having probably been carried off by circulating water. The fossils are reduced to mere internal moulds and the limestone is generally very porous.

Spines of *Cidaris* and *Corals* are frequent. The *Molluscs* being badly preserved, it is impossible to determinate even the genera.

The only determinable species of *Molluscs* were found N. of Redjuno where numerous big shells of *Tridacna gigas* occur together with *Ostrea hyotis, Ostrea lingua, Tridacna rudis, Cassis pila* and *Voluta scapha*.

As *Coral* reefs often vary in thickness, it is no wonder that at some places the limestone is lacking. This appears to be the case in the vicinity of Mt. Pandan near Gajam; to the West of it the first occurrence of Miocene *Coral* limestone of G. Purwala forms an isolated rock with steep slopes.

The *Coral* limestone wholly fails to occur along the southern boundary of Kendeng mountains starting from the valley of river Kedungwarak eastward.

The extensive occurrence of *Coral* limestone on the top of Mt. Seloabang is very remarkable as the dip is very gentle, contrary to the steeply folded older beds N. of it. A similar position was encountered at G. Selubang E. of Dodol. During the fieldwork this observation gave some lead to the supposition of an unconformity existing between the two stages of Miocene beds, but the author refrains from stating this conclusion as proved.

In the Northern anticlines the sole occurrence of *Coral* limestone is found forming the top of G. Sili. A corresponding *Coral* reef occurs (not on this map) on the banks of the Soloriver, near Menden S. of Tjepu, where this river breaks through the last northern anticline of Kendeng mountains.

The occurrence of *Coral* limestone signifies, that in the Upper Miocene the depth of the sea diminished gradually.

*g. Transition beds.*

The uppermost horizon of the Miocene beds is characterized by a change of facies. The Upper Miocene beds generally show a light-gray or white colour with volcanic matter in some horizons but in very small quantities. *Globigerines* and other small *Foraminifera* are generally numerous.

On the other hand the Pliocene beds are of a darkgray or blueish gray colour and abundantly contain volcanic matter. *Globigerines* are scarce and generally small sized.

The wholly different habitus of the Pliocene deposits has to be ascribed to entirely changed conditions of sedimentation.

The extension of the Upper Miocene beds in East and Central Java proves that this area was covered by the sea during the Upper Miocene. The greatest part of Southern Java too was submerged and only here and there small islands or lonely volcanoes reached above the Indian Ocean.

To the North the Ocean was confined by a probably big island, which at present is covered by the Java Sea. From this island the detrital sediments originated which helped to fill the northern part of the already mentioned geosyncline of Northern Java.

These conditions certainly favoured plankton life and the currents from the Ocean caused great amounts of *Globigerines* to be deposited in this submerged part of Java. As volcanism was certainly less active, the deposits mainly consisted of *Globigerina* marls and limestones.

The forming of Coral reefs at the end of the Miocene indicates that the sea gradually lost depth.

In the Pliocene the described conditions ceased to exist.

It is a fact that up till now no marine Pliocene deposits have been discovered in the Southern mountain ranges of Central- and East-Java, neither from the islands of Bali, Lombok, Sumbawa and Flores. The author, who several times crossed the southern mountain ranges in Jogjakarta, Surakarta, Madiun and Kediri, thinks it hardly probably that they have been overlooked. As even the synclinal regions fail to show younger than Miocene marls and Coral limestone, the age of which has been determined for the lower horizons as Tertiary *f*. by the occurrence of *Trybliolepidina Rutteni*, whilst the higher horizons show a typical Upper Miocene habitus by the occurrence of *Globigerina* marls and *Coral* limestone (corresponding to the habitus of the Upper Miocene beds in Kendeng mountains), it is hardly probable that Pliocene marine beds ever were deposited.

Based on these facts we suppose that at the end of the Miocene the Southern mountain range of Java rose above the sealevel and thereby cut off the northern geosynclinal region from the open Ocean.

The remaining narrow sea in northern Java was inclosed between the Southern mountain ranges and the island North of Java (now submerged by the Java Sea). The sediments of this partly closed sea arm show a wholly different character than the sediments of the open Ocean.

The rivers of the expanding southern land brought great quantities of mud, especially as volcanism had become more active on account of the influence of the movements that were cause of the rising action of this southern region. The relics of these volcanoes extend East of Patjitan in Southern Madiun and Kediri.

Sea currents were impeded in this sea arm; no *Globigerines* were brought from the Ocean and only small-sized forms living in the sea arm were deposited. The content of oxygen decreased and was insufficient to oxydize all putrifying organic matter. In this atmosphere the sulphates of the seawater were reduced and by the generating hydrogen-sulphide the ironsulphides were settled that now occur as crystals of pyrite in the Pliocene beds. The dark colour of the beds was caused by unoxydized bitumen.

We are allowed to compare the conditions of the Pliocene sea in this Northern sea arm to that now existing in the street of Madura between Madura and Java and especially as the latter has to be considered a remnant of the former.

The transition beds were formed in the short period when the change gradually took place. Though still containing *Globigerines* in however decreasing numbers, the beds already possess much volcanic matter, causing the sandy habitus and the gray colour. The transition beds are very well developed in the Northern part of Kendeng mountains to a thickness of at least 100 M. in the syncline of Maor and Tretes. They occur in a somewhat reduced thickness near Bareng.

In the Southern Kendeng mountains they seem to be absent between Redjuno and Tjabehan, but as will be epxlained afterwards, this has been caused by a transgression of the Middle Pliocene beds over the Miocene. In the SE. part of the map E. of Kedungwaru river they reappear.

Lower Pliocene.

*h. Darkgray argillaceous layers.*

The Lower Pliocene beds consist of gray or darkgray argillaceous layers, sometimes badly stratified but generally showing thin intercalations of fine sand. The total thickness amounts to

250 M. They are generally barren of fossils. *Globigerines* are small sized and very scarce.

In the Northern Kendeng mountains they occur along the whole border, furthermore in the centre of the syncline of Maor and in the syncline of Tretes. A thin bed of lignite occurs in the region S. of Tondomuljo.

In the Southern Kendeng mountains they are generally absent, due to the above mentioned unconformity of the Middle Pliocene beds. E. of Kedungwaru river, however, where the normal succession of the beds is restored, they reappear.

The Lower Pliocene age of the beds was deduced from their stratigraphical position underlying the Middle Pliocene fossiliferous beds of **Bareng.** In habitus they resemble the lower part of the Lower Pliocene beds in **Sangiran.**

Middle Pliocene.

*i. Conglomeratic limestone, bedded limestone, Balanus limestone.*

To the South of the Kendeng mountains, except in the extreme SE. part of the map E. of Kedungwaru river, the base of the fossiliferous Middle Pliocene (**Sonde**) beds consists of a conglomeratic limestone or a limebearing conglomeratic sandstone directly covering the Miocene *Coral* limestone. The presence of pebbles of Miocene white *Globigerina* marls proves the transgressive character of this basal bed. The conglomeratic habitus is easily recognized in the field on weathered surfaces. The transition beds of the Upper Miocene and the Lower Pliocene argillaceous beds are entirely absent in this region which indicates the presence of a stratigraphic gap.

The latter proves that the southern part of the Kendeng mountains must have been risen above sealevel after the forming of the Upper Miocene *Coral* limestone. This movement corresponds with that of the Southern ranges of Java at the end of the Miocene.

The conglomeratic limestone could be distinctly traced in the field over a distance of several Km. S. of Tjabehan. Owing to bad exposures the investigations S. of Sukun[2] failed to ascertain the presence of this horizon. Farther West, N. of Gajam, however, it was rediscovered as a lime-bearing conglomeratic sandstone in which the matter between the pebbles of *Globigerina* marl consists of volcanic sand.

It appeared that at this spot the directly overlying beds consist of a porous bedded limestone containing small *Foraminifera* and many coarse particles of volcanic matter. Limestones of an entirely identical character occur S. of Sukun at the base of the fossiliferous argillaceous and sandy **(Sonde)** beds. The *Balanus* limestone forms an intermediate layer between the basal conglomeratic limestone and the fossiliferous argillaceous and sandy beds. N. of G. Butak the Pliocene is not well enough exposed to state more than the fact that *Balanus* limestone occurs between the Miocene *Coral* limestone and the volcanic boulder breccia.

The bedded limestone occurs in the bed of the Banjukuning river between Patjar and Gesik and a somewhat resembling limestone in the syncline of Soko is intercalated in the Pliocene beds. Near Gujangan even two beds are present.

To the North of the Kendeng mountains a bedded limestone containing small Foraminifera covers the Lower Pliocene argillaceous beds S. of Geneng. Bedded limestone of a similar character occurs in the syncline of Tretes.

Another bedded limestone was found in river Kalikuntji North of Bara, but it certainly does not represent the same horizon as it is overlying fossiliferous sandy beds.

*j. Mollusc-bearing argillaceous sands. (Bareng and Sonde beds).*

To the North of the Kendeng hills the Lower Pliocene argillaceous beds are directly covered by a series of alternating argillaceous and sandy layers except in the Western part of the map, where the already mentioned bedded limestone occurs as an intermediate layer. The facies is that of a very shallow and muddy sea not far from the shore. As in the preceding period the shore was distinctly more distant, this points to a regression of the sea probably caused by movements in the Kendeng mountains region.

The most important horizons are the sandy and argillaceous beds containing *Molluscs* which were discovered to occur S. of Bareng and which extend eastward to Tondomuljo and westward to the vicinity of Papringan and Ngorogunung. This series of beds generally dips 23—10° N. but it reappears to the North of a syncline between Gebangkerep and Bara, dipping there 8—9° SE—SW. Some horizons of veritable shellbeds occur, but generally the richest fauna was found in argillaceous sandy beds.

The following is a list of the species occurring:

| Marine *Molluscs* from the Middle Pliocene of **Bareng-Turi** (not on the map) and other localities North of Kendeng mountains, partly from a collection made by the author and determined with the assistance of Dr. R. VON KOENIGSWALD and partly from a still unpublished collection determined by Dr. I. M. VAN DER VLERK | Miocene | Pliocene | Recent | **Tjilanang** Upper Miocene | **Tjidjurai** Middle Pliocene | **Sonde** Middle Pliocene |
|---|---|---|---|---|---|---|
| *Gasteropods* | | | | | | |
| Terebra talahabensis Mart. [1]. . . | + | — | — | | | |
| Conus vimineus Reeve [1] [2]. . . . | — | + | + | | | + |
| „ socialis Mart. [2] . . . . . | + | + | — | | | + |
| „ scalaris Mart. [1]. . . . . . | + | — | — | | | |
| „ ngavianus Mart. [1] . . . . | + | + | — | | | + |
| Surcula nodifera Lam. var. Woodwardi Mart. [1] [2] . . . . . . . | + | + | + | | + | + |
| Pleurotoma carinata Gray var. [1] . | + | + | + | | + | + |
| „ gendinganensis Mart. [1] [2]. . . . . . | — | + | — | + | + | + |
| Drillia flavidula Lam. [1] [2] . . . . | — | + | + | | | + |
| Cancellaria asperella Lam. [1] [2] . . | — | + | + | | | + |
| Trigonostoma crispata Sow. [1] [2] . | — | + | + | | + | |
| Olivancillaria subulata Lam. [1] . . | + | + | + | + | | |
| „ gibbosa Born. [1] . . | + | — | — | + | | |
| Marginella quinqueplicata Lamk. [1] [2] . . . . . | — | + | + | + | + | + |
| „ quinqueplicata var. min. Mart. [1] [2] . . . . . . . | + | + | + | + | + | |
| „ dactylus Lamk. [1] [2] . . | + | + | + | | + | |
| Cymbium Lamarckii Mass. [2] . . | — | — | + | | | |
| Voluta Junghuhni Mart. [1] . . . . | + | — | — | + | | |
| Mitra adusta Lam. [1] . . . . . . | — | + | + | | | + |
| „ flammea Quoy [1] . . . . . | + | + | + | | | + |
| Fusus menengtenganus Mart. [1]. . | — | + | — | | | |
| Clavilithes Verbeeki Mart. [1] [2] . . | + | + | — | + | | + |
| „ Fennemai Mart. [1]. . . | + | — | — | | | |
| Cominella sangiranensis Mart. [1] [2] | — | + | — | | + | |
| Lathyrus losariensis Mart. [1] . . . | — | + | — | | | |
| Melongena cochlidum Linn. [1] [2] . | + | + | + | + | | |
| „ madjalengkensis Mart. [1] . | — | + | — | | + | |
| Dipsaccus pangkaënsis Mart. [1] [2] . | + | + | — | | | |
| „ gracilis Mart. [1] [2] . . . | + | + | — | + | | |
| Nassa Verbeeki Mart. [1] [2] . . . . | + | + | — | | | + |
| „ gemmulata Lamk. [1] [2]. . . | — | + | + | | | |
| „ coronata Brug. [1] . . . . . | — | + | + | | | + |
| Murex Verbeeki Mart. [1] . . . . . | + | + | — | + | | + |
| „ ejectus Mart. [1] [2]. . . . . | — | + | — | | + | |
| „ djarianensis Mart. [2] . . . | + | — | — | + | + | |
| Purpura carinifera Lamk. [1] . . . | — | + | + | | + | |
| Eutritonium pileare Linn. [1] . . . | + | + | + | + | + | |
| „ pseudopyrum Mart. [2] | + | + | — | + | | |
| Hindsia tjemoroënsis Mart. [1]. . . | ? | + | — | | | |
| Ranella subgranosa Beck. [1] [2] . . | + | + | + | | + | + |
| „ nobilis Reeve [1] [2] . . . . | + | + | + | + | | + |

[1]) collected by the author.

[2]) determined by Dr. VAN DER VLERK.

| Marine *Molluscs* from the Middle Pliocene of **Bareng-Turi** (not on the map) and other localities North of Kendeng mountains, partly from a collection made by the author and determined with the assistance of Dr. R. VON KOENIGSWALD and partly from a still unpublished collection determined by Dr. I. M. VAN DER VLERK | Miocene | Pliocene | Recent | **Tjilanang** Upper Miocene | **Tjidjurai** Middle Pliocene | **Sonde** Middle Pliocene |
|---|---|---|---|---|---|---|
| Ranella bitubercularis Lamk. [1] [2] | + | + | + | | | + |
| „ pulchra Gray [1] . . . . . | + | + | + | | | |
| „ magnifica Mart. [2] . . . . | + | — | — | ? | | |
| Siphonalia paradoxica Jenk. [1] . . | + | + | + | | + | |
| Cassis pila Reeve . . . . . . . . | + | + | + | ? | + | + |
| Dolium costatum Desh. [1] . . . . | + | + | + | | | + |
| „ modjokasriense Mart. [2] . | ? | + | — | | | |
| „ zonatum Green [1] [2] . . . | — | + | + | | + | |
| Cypraea sondeiana Mart. [1] [2]. . . | + | + | — | | | + |
| Strombus isabella Lamk. [1] [2]. . . | — | + | + | | | + |
| „ Fennemai Mart. [1] . . . | — | + | — | | | + |
| „ madiunensis Mart. [2] . . | — | + | — | | | + |
| Potamides cheribonensis Mart. [1] . | — | + | — | | + | |
| „ sulcatus Brug. [1] . . . | + | + | + | + | | |
| „ palustris [2] . . . . . . | + | + | + | | | |
| Telescopium telescopium Linn. [2] . | + | + | + | + | | |
| „ titan Mart. [1] . . . . | — | + | — | | + | |
| Vermetus javanus Mart. [1] . . . . | + | + | + | + | + | + |
| Turritella bantamensis Mart. [1] . . | + | + | — | | | |
| „ javana Mart. [1] . . . . | + | + | — | + | | |
| „ vittulata Reeve [1] . . . | — | + | + | | | |
| „ djadjariensis Mart. [1] . | + | + | — | + | + | |
| „ terebra Lamk. [1]. . . . | — | + | + | | | |
| Solarium perspectivum Linn. [1] [2] . | + | + | + | + | + | + |
| Solarium maximum Philippi. [2] . . | + | + | + | | | + |
| Natica ala-papilionis Chemn. [1] . . | — | + | + | | | |
| „ globosa Chemn. [1] . . . . | + | + | + | + | + | |
| „ rufa Born. [1]. . . . . . . | + | + | + | + | | + |
| „ melanostoma Gmel. [1]. . . | + | + | + | + | | + |
| „ lineata Lamk. [1] . . . . . | + | + | + | | | |
| „ mamilla Linn. [2] . . . . . | + | + | + | + | | + |
| „ powisiana Recluz. [2] . . . | + | + | + | + | | |
| Crucibulum extinctorium Lamk. [1] [2] | — | + | + | | | |
| Eulima sondeiana Mart. [1] . . . . | — | + | — | | | + |
| *Lamellibranchiates.* | | | | | | |
| Ostrea disciformis Mart. [1] . . . . | + | + | — | + | + | + |
| „ lingua Sow [1] [2]. . . . . . | + | + | — | + | + | |
| Placuna placenta [1] . . . . . . . | + | + | + | | + | |
| „ pseudoplacenta Mart. [1] [2] | — | + | — | | | |
| Anomia Boettgeri Mart. [1] . . . . | — | + | — | | | |
| Plicatula imbricata Menke. [1]. . . | — | — | + | | | |
| Amussium Hulshofi Mart. . . . . | + | — | — | | | |
| Chlamys senatorius Gmel. [1] [2] . . | + | + | + | + | + | + |
| „ kebolintangensis Mart. [1] | — | + | — | | | + |
| „ tjaringinensis Mart. [1] . . | — | + | — | | | |

[1]) collected by the author.
[2]) determined by Dr. VAN DER VLERK.

| Marine *Molluscs* from the Middle Pliocene of **Bareng-Turi** (not on the map) and other localities North of Kendeng mountains, partly from a collection made by the author and determined with the assistance of Dr. R. VON KOENIGSWALD and partly from a still unpublished collection determined by Dr. I. M. VAN DER VLERK | Miocene | Pliocene | Recent | **Tjilanang** Upper Miocene | **Tjidjurai** Middle Pliocene | **Sonde** Middle Pliocene |
|---|---|---|---|---|---|---|
| Pinna vexillum Born. [1]: | + | + | + | + | + | |
| Arca bistrigata Dkr. [1] | — | + | — | | + | |
| „ oblonga Phil. [1] | + | + | + | + | | |
| „ inaequivalvis Brug. [1] | — | + | + | | + | |
| „ tambacana Mart. [1] | + | + | — | | + | |
| „ gendinganensis Mart. [1] | — | + | — | | | + |
| „ rustica Mart. [1] | — | + | — | | | |
| „ ferruginea Reeve [2] | — | + | + | | | |
| „ palabuanensis Mart. [2] | + | — | — | | | |
| „ scapha Chemn. [1] | — | + | + | | | + |
| Limopsis venusta Mart. [1] | ? | + | — | | [3] | ? |
| Crassatella radiata Sow. [1] | — | + | + | | | |
| Dosinia boettgeri Mart. [1] [2] | + | + | — | + | + | |
| „ juvenis Chemn. [2] | + | + | + | | | |
| Venus chlorotica Phil. [1] [2] | + | + | + | + | + | |
| Cyrena rustica Mart. [1] | + | + | — | | [3] | |
| Tellina Dyki Mart. [2] | — | + | — | | | |
| Corbula socialis Mart. [1] | + | + | — | + | + | |
| „ scaphoïdes Hind. [1] | — | + | + | | | |
| Pholas hercules Mart. [1] | — | + | — | | + | |
| Total 104 species | 60 | 94 | 55 | 33+2 | 37 | 36 |

The list contains 104 species, 55 or 53% of which are still living forms. Several species remained undetermined.

No doubt this fauna has to be regarded as a Middle Pliocene one. The percentage of still living species quite corresponds with that of other known occurrences of Pliocene age e.g. **Tjidjurai** 51%, **Sonde** 53%. Moreover only 9 species known as yet to occur exclusively in Miocene beds are present, whilst 41 so far have always been found restricted to Pliocene or Pliocene-Recent beds and 2 to the times being.

The Upper Miocene **Tjilanang** beds show 33 + 2 identical forms against 37 and 36 occurring in the Pliocene of respectively **Tjidjurai** and **Sonde.**

Though the numbers of corresponding fossils are almost equal, the resemblance to the **Tjidjurai** fauna is considerably greater than to **Sonde** and **Tjilanang,** the number of compared species

[1] collected by the author.
[2] determined by Dr. VAN DER VLERK.
[3] collected from Tjidjurai and determined by the author.

from these three spots being respectively 66, against 150 and 189. This signifies that 57½% of the **Tjidjurai** Molluscs against only 24% of the **Sonde** fossils and 18½% of the **Tjilanang** fossils are identical to the **Bareng** fauna.

Special mention must be made of a femur of *Rhinoceros* discovered by the author in the marine fossiliferous beds N. of Bara.

To the South of the Kendeng mountains the Middle Pliocene beds are of a different character. The thickness of the marine beds is nowhere more thans 100 M. and near Tjabehan it even does not pass 15 M.

The basal bed is the already mentioned conglomeratic limestone or a limebearing conglomeratic sandstone containing pebbles of Miocene white *Globigerina* marl, the latter pointing out its transgressive character.

The argillaceous and sandy beds forming the main fossiliferous horizon much resemble the **Sonde** beds from the **Trinil** region and almost certainly represent the same horizon, as the outcrop, with only one interruption near Ngawi, could be traced from **Sonde** to the region under discussion.

Though *Molluscs* were collected, the total number appeared to be too small for a reliable comparison.

| List of *Molluscs* from the Middle Pliocene argillaceous sands near **Sukun, Soko** and **Tjabehan,** collected by the author and determined by Dr. R. VON KOENIGSWALD | Upper Miocene | Pliocene | Recent | **Sonde** | **Padas-malang** |
|---|---|---|---|---|---|
| Marginella dactylus Lamk.. . . . | + | + | + | | + |
| „ quinqueplicata Lamk. . | + | + | + | + | |
| Cominella sangiranensis Mart. . . | — | + | — | | |
| Dipsaccus gracilis Mart. . . . . | + | + | — | | |
| Cassis pila Reeve . . . . . . . . | ? | + | + | + | |
| Ostrea hyotis Linn.. . . . . . . | + | + | + | + | |
| Placuna placenta Linn. . . . . . | + | + | + | | |
| Chlamys senatorius Gmel . . . . | + | + | + | + | |
| Cyrena rustica Mart. . . . . . . | + | + | ? | + | |
| Corbula socialis Mart.. . . . . . | + | + | — | | |
| „ scaphoides Hinds . . . | — | + | + | | |
| Total number of species 11 . . . | 8 + 1 | 11 | 7 + 1 | 5 | 1 |

Several species not determined.

MARTIN (1926), when comparing the fauna of **Tjidjurai** and **Sonde,** concluded to a probable slight difference in age. The resemblance of **Bareng** and **Tjidjurai** implies that the same question arises with regard to the differing faunas of **Bareng** and **Sonde.** There certainly is a difference in facies. Whilst in **Bareng** the sea seems to have been mud laden, the **Sonde** beds often contain many *Corals* proving the existance of a somewhat clearer sea. This difference in facies probably was caused by reverse tilting movements. In **Bareng** the sequence of the beds obviously indicates a regression of the sea, whilst on the southern border of the Kendeng mountains the Middle Pliocene **Sonde** beds point out a transgression. During a regression, of course, erosion in the rising land is strong. The rejuvenated rivers are able to cut into the deposits of the old courses. They possess a great difference in level and are able to bring immense quantities of mud into the sea. Deltas soon expand and the shore is rapidly brought forward.

On the other hand a transgression means, that erosion is impeded in the sinking land. The mouths of rivers are flooded and the small quantities of mud transported by the rivers settle in the aestuaries. The sea is clear of mud, favouring the growth of *Corals*.

This reasoning leads to the conclusion that tilting movements of reverse sense would explain both the facial and faunistic differences of **Bareng** and **Sonde.**

A difference in age however is highly improbable, the percentage of living species of Molluscs being the same (53%).

Upper Pliocene.

*k. Upper argillaceous beds.*

To the N. of the Kendeng mountains the **Bareng** beds are covered by argillaceous beds still containing marine *Molluscs*. S. of Kalikuntji a bed of *Ostrea* shells occurs, whilst W. of Babad a *Mollusc*-bearing bed was discovered, built up for the greater part by volcanic matter e.g. pumice stone, which proves that it was deposited contemporaneously with heavy volcanic eruptions. The Upper Pliocene marine beds in the syncline N. of Bareng being unconformably covered by Pleistocene *Vertebrate*-bearing river deposits, there is no certainty that the highest horizons of the marine Pliocene are present.

The total thickness of the Lower, Middle and Upper Pliocene beds N. of the Kendeng mountains is estimated by the author to amount to 700 M. A borehole 7 Km. E. of Bodjonegoro drilled to a depth of 435 M. by the „Dienst van het Grondpeilwezen" did not reach the base of the marine Pliocene beds.

In the Southern Kendeng mountains the upper argillaceous beds are restricted to an small patch in the centre of the syncline of Soko.

1. *Corbicula and Melania sandstone, volcanic breccia, tuffs, sandstones and conglomerates (Butak beds).*

All along the southern border of the Kendeng hills the marine Pliocene beds (**Sonde** beds) are covered by a mighty complex of terrestrial volcanic sediments. Several beds of volcanic boulder breccia alternate with tuffs, sandstones, conglomerates and *Corbicula* or *Melania* sandstone.

Some of the beds of volcanic boulder breccia are certainly contemporaneous with the marine beds. South of **Sukun** two beds of volcanic breccia are intercalated in the Middle Pliocene marine beds and contain marine fossils.

The lowest horizon of volcanic breccia N. of Redjuno likewise contains marine fossils, but a marine origin of this bed is not provable as the fossils may have been derived from the underlying beds during the origin of the boulder breccia by the strong erosive action of „lahars".

In the remarkable syncline of **Soko** a bed of volcanic breccia, containing fresh-water *Melania* is covered by marine argillaceous beds.

This shows the volcanic breccia of Soko to be intercalated in the marine beds. As no doubt this bed is identical with the lowest mighty volcanic boulder breccia South of the line Padjeng-Tjabehan — both occupying a corresponding position on opposite slopes of an anticline — it is obvious that the lower part of the terrestrial volcanic beds along the southern border of the Kendeng mountains must be contemporaneous to marine Pliocene beds. This explains the sometimes extremely volcanic habitus of some Upper Pliocene marine beds near Bara N. of the Kendeng mountains. Also the extreme thinness of the marine Pliocene beds S. of the Kendeng mountains — near Tjabehan even reduced to 15 M. — is now easily understood.

The Upper Pliocene and perhaps part of the Middle Pliocene being represented S. of Kendeng mountains by terrestrial volcanic sediments, those 15 M. of marine sediments only correspond with the lower part of the 450 M. marine Middle and Upper Pliocene beds N. of the Kendeng mountains.

South of Sukun the series of marine Middle Pliocene beds with intercalations of volcanic breccia, is covered by freshwater beds consisting of tuff with several species of *Melania* and *Isidora sp.* and a limebearing sandstone with *Corbicula sp.* The superincumbent bed is a volcanic breccia.

N. of Bangle and Sumberkepuh fresh-water beds, containing several species of *Melania* alternate with *Vertebrate*-bearing sandstones. They occupy a corresponding position between marine argillaceous beds and volcanic breccia. Remains of *Vertebrates* were discovered by the author in conglomerates and sandstones S. of Ngluju. The occurrence of *Vertebrates* just S. of Mt. Butak was discovered by DUBOIS (1907), who made some excavations at this spot in 1893, and has been described by VERBEEK (1908).

The author collected from this spot some teeth of *Crocodilus* and some remains of *Stegodon sp.* The *Vertebrates* occur in a tuff, containing small lumps of pumice stone, the tuffs and the volcanic boulder breccia originating from Mt. Wilis. This is easily understood for the volcanic breccia occurring E. of Tjabehan, considering the increasing distance from the so much smaller volcano Mt. Pandan, though it is not entirely improbable that the latter too produced some of the products in its neighbourhood.

Especially with regard to an andesitic lava flow between Gajam and Krondonan, intercalated in steeply folded volcanic beds and an occurrence of lava in the valley of R. Genuk N. of Tritik, the author inclines to admit an origin from Mt. Pandan.

The terrestrial volcanic sediments at the base have been proved to be contemporaneous to marine Pliocene beds, but near the top they possibly are already of Pleistocene age. It is not inconceivable that the highest horizon of volcanic boulder breccia corresponds with a Pleistocene boulder breccia which overlies unconformably the Pliocene and Miocene marine beds in the region between Ngawi and Redjuno.

In the Mt. Pandan region, however, no traces of a hiatus or an unconformity are visible; the intercalation of volcanic breccia in marine Pliocene beds on the contrary proves an unbroken succession of sediments.

An important observation is the occurrence of fossil *Vertebrates* in beds just overlying the lowest bed of volcanic breccia, a Pliocene age of which is certain. Obviously thus in this region remains of *Vertebrates* are not restricted to the Pleistocene only, but may occur as well in the Pliocene (probably Upper Pliocene.)

Pleistocene.

*m. Sandstones, conglomerates and black clay with Vertebrates (Trinil beds).*

N. of the Kendeng mountains the beds corresponding with the *Vertebrate* horizon of **Trinil** show to have been unconformably deposited upon the folded marine tertiary beds.

They as well cover the Upper, Middle and Lower Pliocene and even the Miocene beds. Erosion separated the deposits into smaller or larger patches. As there are signs of a feeble posthumous folding or tilting movement, the largest patches persisted in the neighbourhood of the synclinal axis of the Pliocene beds N. of Bareng-Tondomuljo.

The **Trinil** beds are very well developed in the valley of River Kuntji near campong Kalikuntji. The basal bed mainly contains volcanic sand and andesitic pebbles together with thickshelled marine fossils showing much wear and having certainly been washed down from eroded Pliocene beds. A somewhat higher sandy horizon contains lime concretions with imprints of fossils derived from the Pliocene beds and pebbles of white *Globigerina* marl. The mainly volcanic sand often is found to contain *Globigerines*, but these of course represent detritus of Miocene beds, as there cannot exist any doubt concerning the terrestrial and fluviatile origin of the deposits. The occurrence of *Unio sp.* proves the fresh-water origin.

To the South of the Kendeng mountains the outcrop of **Trinil** beds extends all along the border of the range, except just South of Mt. Pandan where the outcrop is covered by younger products of Mt. Wilis and Mt. Pandan. The thickness of this series increases from W. to E. and is very great near **Tritik.**

The **Trinil** beds are best characterized as river deposits, being mainly sandstones and conglomerates, alternating with tuffs and beds of dark clay. Fresh-water fossils often abound and form distinguishable horizons, 5 of which were found near Sumberbenda.

In some cases there is no doubt about the contemporaneousness of the *Vertebrates* with the deposits in which their remains have been buried. The good preservation especially of those occurring in tuff or fine-grained beds gives adequate evidence.

In other instances it is quite impossible to admit that an already fossilized and therefore weighty bone or a molar should occur in a deposit consisting of mere sand in which pebbles are entirely absent. Especially with a view to the weighty skulls and lower jaws of big vertebrates it is quite improbable that these should have been derived from older beds. In many cases adjoining parts of the same specimen were found close to one another in such a way that even transport is excluded.

The pebbles of the conglomerates mainly are of volcanic origin but they are sometimes mixed with detritus of Miocene marls and limestones. This especially seems to have been the case in the neighbourhood of Kedungbrubus where VERBEEK (1908) stated the occurrence of *Globigerines*, which caused a fully wrong conception about the age of the beds, called Miocene, whereas in reality the *Globigerines* only join the detritus of Miocene beds.

The amount of material washed down from the Miocene beds often implies a considerable content of lime which solidifies the conglomerates or concentrates in concretions.

Several species of *Melania* and *Unio* occur.

In weathering, the conglomerates and sandstones lose their content of lime, and form exposures of loose gravel and sand.

A great deal of the *Vertebrates* collected by DUBOIS originates from the **Trinil** beds to the South of Kendeng mountains, more closely from **Redjuno, Kedungbrubus, Tritik** and **Bangle.**

*n. Coarse and fine grained tuffs. (Notopuro beds).*

The overlying beds consist of a monotonous succession of fine and coarse tuffs, sandstones and conglomerates in which no *Vertebrate* remains seem to have been deposited.

In a Southward direction they seem to pass gradually into the slopes of the Wilis-volcano. The slight dip to the S., generally not surpassing 3°, proves that they still have been influenced by tilting movements.

o. *Volcanic breccia and tuffs.* (*Pandan Volcano*).

The survey in the vicinity of Mt. Pandan clearly proved that the last eruptive products of Mt. Pandan are younger than the **Trinil** beds. S. of Mt. Pandan the latter are only exposed where the Pandan tuffs were carried away by erosion.

The relatively young age of Mt. Pandan is also evident to the N of it where the folded and mainly eroded Miocene is covered by volcanic products. It must be admitted however, that volcanic boulder breccia and conglomerates found in or on the slopes of the river valleys do not necessarily originate from volcanic eruptions. It is even probable that some of the boulder breccia originates by landslides. The habitus is different as fine tuffoid or sandy matter is nearly absent and the deposit mainly consists of an accumulation of loose big boulders.

A boulder breccia of this kind could be traced over a great distance in the vicinity of Tadahan. Since its origin the river Gandong has been able to cut itself down to a 20 M. deeper bed in the Miocene marls. The landslide therefore dates from a rather distant time and the last eruption being still earlier, it certainly cannot date from Holocene time. The age of the last eruptions is supposed to be the same as that of the **Notopuro** beds.

Some of the highest peaks of Mt. Pandan are built of solid andesite. On the northern slopes some „bocca's" protrude probably likewise consisting of andesite. The springs in campong Kedaton originate from a lava flow on the western slopes.

To the North of Mt. Pandan the volcanic products extend far northward and several dykes jut out forming separate centres of eruption. The most marked of these tops is G. Lawang.

The tectonic structure of the Kendeng mountains shows the existence of a remarkable transverse fault. Volcanism is quite extinct now but posthumous activity is shown by some hot springs. One of these springs occurs S. of Tengaring, another one N. of Djomblang and a third one S. of Djari. The latter is very remarkable by its fossil deposits of calcareous sinter. The tempara-ture is 35° C. and no sinter is formed in recent times.

The chemical content of the water of the two last-mentioned springs has been analysed in the Laboratory of the „Proefstation voor Waterzuivering" at Mr. Cornelis.

| | Djari: | | Djomblang: | |
|---|---|---|---|---|
| Dried rest . . . | 6203 | mgr./L. | 1292 | mgr./L. |
| Glowed rest . . | 5923 | „ | 1102 | „ |
| Ca . . . . . . . | 244 | „ | | |
| Mg . . . . . . . | 22,5 | „ | | |
| Fe . . . . . . . | 1,3 | „ | 0,1 | „ |
| Mn . . . . . . . . | 1.32 | „ | | |
| $HCO_3$ . . . . | 1750.7 | „ | | |
| $SiO_2$ . . . . . | 213 | „ | 142,5 | „ |
| $SO_4$ . . . . . . | much | | present | |
| Cl . . . . . . . | 2449.5 | „ | 276 | „ |
| Hardness . . . | 39 | D° | 4.5 D° | |

*Structure.*

The structure of the Kendeng mountains is best illustrated by the map and the sections.

*Section a-b-c-d-e. Kali Kedungwarak-Gampeng-Puntju-Kali Djomblang.*

The thrustplane, marking the northern border of the imbricate structure of the Kendeng mountains appeared to be very distincly exposed in river Borik S. of Bareng. It was found to dip 30° S. and showed that the rather strongly compressed gray *Globigerina* marls were thrusted over the gently northward dipping white *Globigerina* marls. It must be mentioned that the field observations exclude any other explanation of the established facts.

At the back of the two northernmost anticlinal folds, the structure is rather complicated. Two other thrustplanes were recorded in the field and the interjacent synclinal region is strongly compressed. The third anticline shows signs of backfolding.

This group of three anticlines is separated from a southern group by a broad syncline in which the *Globigerina* limestone topped by small patches of Coral limestone is only gently dipping.

In this section one is lead to the conception, that the gray *Globigerina* marls from the Northern anticlines correspond with white *Globigerina* marls in the Southern region. The author admits, however, that there might be some possibility of the first being partly contemporaneous to the upper part of the alternating beds of tuffs and marls.

The southern anticlines do not show a complex structure. This however, only holds true for the S.E. part of the investigated region as the other sections in the vicinity of Mt. Pandan prove the existance of upthrusts.

The Southern border of the Kendeng mountains is marked by an unconformity between Middle Pliocene Sonde beds and Miocene beds.

*Section f-g. Kali Kuntji.*

This section was meant to indicate the unconformable position of the Trinil beds N. of the Kendeng mountains on top of Pliocene beds of different horizons. Observations W. of Babad, however, are still more interesting. Several signs indicate the existance of a much disturbed region.

In a shallow pit dug by the natives a thrustplane was visible with a dip of 40° S., Pliocene beds covering Trinil beds. Some 50 M. W. of this pit steep dips and a very intricate position of marine Pliocene beds and Trinil beds were observed along the road. This line of disturbance coincides conspicuously with the topographic features of the region, as just along it the hilly country breaks off abruptly towards the lower plain N. of it.

The author inclines to accept a post-tertiary upthrust movement for the explanation of the described phenomena. This movement on a smaller scale resembles the upthrust movement along the northern border of the Kendeng mountains which is due to the Pliocene folding.

*Section h-k. Mengkal-Tjabehan-Soko.*

This section intends to show the position of the lowest bed of volcanic boulder breccia.

South of an anticline with a core of Miocene *Coral* limestone the volcanic breccia is covered by terrestrial deposits, whereas N. of it the superincumbent bed in the syncline of Soko was found to consist of marine clay. The breccia thus is intercalated between the marine beds and the covering terrestrial beds of the southern region are partly contemporaneous to these marine sediments.

It is to be noted that the southern slope of the anticline dips certainly stronger than the northern slopes; this indicates a tendency of backfolding.

The unconformable position of the Sonde beds is evident.

*Section l-m-n-o. Tritik-Sukun-Maor.*

The *imbricate* structure of the northern anticlines was distinctly proved in the field. The thrustplane of the second anticline was visible in the region of campong Gedangan, dipping about 40° S. In the northern anticline south dipping Miocene gray *Globigerina* marls touch north dipping Lower Pliocene dark gray clays.

The syncline of Maor is very interesting, a patch of Lower Pliocene beds being saved here from erosion by its position in the centre of the syncline.

In the southern anticlines an *imbricate* structure was likewise found to exist. The succession of the beds, however, is somewhat unusual as the *Coral* limestone seems to cover directly the much older alternating tuffs and marls. The author refrains from anticipating an unconformity between the two horizons of the Miocene, though he admits that other observations in the neighbourhood of Dodol produce a similar impression.

Intercalations of volcanic breccia in the Sonde beds prove the increase of volcanism. The overlying beds are caracterized as terrestrial and lacustrine deposits by the occurrence of *Corbicula* and *Melania* sandstone. After the observated dips, the Trinil beds near Tritik must be of extremely great thickness.

*Section p-q-r. Notopuro-Kedungbrubus-Butak-Kalimas.*

By means of this section the author intends to correct the section published by VERBEEK (1908). The beds occurring between G. Butak and Notopuro were formerly erroneously supposed to embrace Lower Miocene breccias, Middle Miocene marls and Pleistocene conglomeratic and sandy layers. In reality, however, the volcanic breccias alternating with tuffs sandstones and conglomerates are of Upper Pliocene age. The vertebrate horizon of G. Butak is intercalated between two beds of volcanic breccia, and belongs to a deeper level than the Pleistocene **Trinil** beds which are exposed in the visinity of Kedungbrubus. The uppermost horizons called **Notopuro** beds are to be considered as Upper Pleistocene.

Marine beds are present N. of Mt. Butak only. The *Balanus*-limestone of Middle Pliocene age occurs between the lowest bed of volcanic breccia and Miocene *Coral* limestone. In the Miocene beds two anticlines were encountered both showing an upthrust in the Northern slope.

*Section s-t-u-v-w. Redjuno-Kali Gedong.*

This section shows a similar structure of the two anticlines which are represented in it. The lowest bed of volcanic breccia appeared to contain marine fossils, but there is no certainty as to its being a marine deposit.

*Section x-y. Sumberbendo-G. Berdjo.*

In this section an idea is given of the habitus and the thickness of the layers belonging to the **Butak** and **Trinil** beds. The total thickness exceeds 600 M.

All sections point out that the structure of the Kendeng mountains is governed by *imbrication.* This structure has been caused by the yielding of the northern slopes under an increasing thrust from the South. Towards the East the folding thrust certainly was less heavy. In the SE. part of the map at least two anticlines show a regular structure. Likewise the imbricate structure of the Northern anticlines decreases in an Eastern direction. Though signs of it are still visible in the region N. of Djombang the upthrusts certainly diminish and even become absent farther Eastward Moreover the anticlines divergate in an Eastern direction and — though partly outside the accompanying map — it appears that they possess a distinct Eastward plunge, which causes them to disappear one after the other, starting with the southernmost one.

All these facts state that the intensity of the folding movements decreased in an Eastern direction.

It is to be noted, however, that the disappearance of most anticlines is apparent only. It seems that in the plain of Modjokerto and Djombang several culminations — though indicating less strong folding movements — exist.

In the vicinity of Mt. Pandan the trend of the anticlines is quite abnormal. N. and NW. of this mountain a NW.—SE. direction is dominant, whereas NE. of it a NE.—SW. trend is observed.

RUTTEN (1927) supposed Mt. Pandan — presumed to be a older andesitic body — to be the cause of this deviation of the structural directions. However, it has already been pointed out, that the first activity of Mt. Pandan certainly took not place before the Middle or Upper Pliocene. The author thinks it more likely that the deviation was due to the presence of an important transverse fault, which moreover favoured volcanic activity in this region. Similar deviations of folding axisses occur in several parts of Java e.g. in southern Tegal where andesitic bodies —

which according to RUTTEN'S hypothesis might have caused the deviation — are missing.

The upthrust-plane indicating the Northern border of the Kendeng mountains was recorded in all places where investigations have been made. N. of Tretes, though no own fieldwork was done, its existance is known to the author from still unpublished observations of other geologists.

The transverse and oblique faults given on the map, were either directly observed during the field work and in that case indicated as fully proved, or deduced from observations as to otherwise unexplainable changes of dip and habitus of the beds of neighbouring areas.

This faulting of course is a phenomenon accompanying the folding movements, whereby local differences of thrust are compensated.

It appears that the oblique faults of a SE.—NW. direction are very important as they often show a considerable horizontal displacement besides a possible vertical one. Even the plane of the upthrust movements along the Nothern border of Kendeng mountains was influenced by them, as is appearing from its broken line on the map.

The unconformable position of the Pleistocene *Vertebrate* beds N. of Kendeng mountains proves that the main folding of the tertiary beds took place during the Upper Pliocene.

Evidently earlier movements caused the upheaval of the Upper Miocene coral reefs in the Southern part of Kendeng mountains, which region was invaded again by the sea at the beginning of the Middle Pliocene.

On the other hand a slight upthrust of Pliocene over Pleistocene beds in the vicinity of Bara N. of Kendeng mountains and the slight dip of the Pleistocene beds proves that some movements were still going on in Pleistocene time. The folding certainly had its maximum during the Upper Pliocene and was accompanied by a strong volcanic activity.

### 13. **Kendeng mountains N. of Djombang.**

(*With one map and 8 sections*).

RUTTEN (1916) published a section through the hillrange between Ngimbang and Kabuh (part of which corresponds with our section A'—B') in which two anticlines were represented, the southern one showing signs of an upthrust in the northern slope. The sediments were divided in a lower fine grained stage of badly stratified blueish-gray, argillaceous *Globigerina* marls of Miocene age and an upper coarse grained stage of sandy marls, sandstones, conglomerates and volcanic tuffs, embracing beds of younger age including the Trinil beds.

KOCH (1924) published the result of palaeontologic researches concerning a fauna of smaller *Foraminifera* originating from a specimen of marl collected in 1905 by Dr. M. MÜHLBERG on the road between Babad and Djombang N. of **Kabuh** and probably representing a horizon of the above mentioned lower stage of *Globigerina* marls.

The fauna appeared to consist of 107 species of *Lituolidae, Miliolidae, Lagenidae, Textularidae, Globigerinidae, Rotalidae* and *Nummulinidae*. Many of these were found to be extinct, but identical or related to forms described by C. SCHWAGER from Pliocene argillaceous beds of the Nicobar Islands, Taking into consideration that of 54 species of *Foraminifera* described by P. J. FISCHER from Seran, only a few were found to be extinct, though the age was proved to be Pliocene by means of an accompanying rich fauna of *Molluscs*, KOCH for the bed of **Kabuh,** concluded a Lower Pliocene or a Miocene age.

Two of the described species viz. *Uvigerina javana Koch* and a new species of *Globigerina* were considered to be serviceable as guide fossils.

In 1927 the author collected some data regarding the habitus and the dip of the beds and the occurrence of fossil vertebrates. Dr. J. M. W. NASH, who in 1930—1931 collaborated with the author in surveying the artesian basin extending from Northern Madiun to Southern Surabaja, kindly allowed the author to make use of the result of his investigations which for the greater part form the base of the following description.

*Situation*: The mapped region embraces part of Kendeng mountains N. of Djombang where the height is reduced to less than 200 M. above sealevel.

*Stratigraphy.*

Upper Miocene.

*a. White-yellow, thinly bedded Diatom marls.*

The lowest horizon exposed in Kendeng mountains N. of Djombang consists in an over 475 M. thick complex of tuffaceous sandy marls containing *Globigerines* and *Diatoms*. It is well exposed in the Section near Gesing (A-B-C). In other sections the facies corresponds, but generally a smaller part of this horizon is exposed.

The colour of the beds generally is creamy white. The fine-grained marls are easily powdered by a slight pressure (between thumb and forefinger). Under the microscope an argillaceous character is evident. Tuffaceous matter is present. The *Diatom* beds possess a low specific weight and easily absorb water, two features by which they are easily recognized in the field as the *Diatoms* only show under the microscope when 100 times magnified.

In the well stratified *Diatom* marls some intercalations occur of more compact *Globigerina* marls to a maximum thickness of 1 M. In these beds where *Globigerines* prevail, the number of *Diatoms* is much less. An interbedded sandy clay of 40—45 cM. thickness, void of fossils, occurs in the section near G. Dakon (A′—B′).

This horizon of *Diatom-Globigerina* marls corresponds with the upper part of the breccias, tuffs, sandstones and marls of the Mt. Pandan region, where *Diatom* beds already have been described to occur in the region N. of Sendanggogor.

*b. White-yellow Globigerina marls (top-beds blue-gray).*

The superincumbent beds of a thickness varying between 180 and 225 M. consist of *Globigerina* marls in which *Diatoms* are absent. They are generally badly stratified and compact. The lower beds are likewise of a creamy white colour but the upper horizons possess a smudgy gray to blueish gray colour. In those upper horizons meanwhile the number and size of the *Globigerines* decreases. Whilst the lower horizons correspond with the white *Globigerina* marls of the Mt. Pandan region, the highest beds are to be considered as representing the transition beds between the Miocene and the Pliocene.

An intercalation of *Mollusc*-bearing limestone to a thickness of 5 M. occurs S. of Gesing (Section A—B) just between the two types of *Globigerina* marls. There is some possibility of this bed representing the same stratigraphic level as the *Coral* limestone in the Mt. Pandan region.

Though the habitus of the beds not in all features resembles that of the Kendeng mountains in the vicinity of Mt. Pandan, the succession of almost identical beds shows that the same division of the Upper Miocene is present in both regions. An Upper Miocene age is not contradictory to the results of the palaontological investigations by KOCH.

Lower Pliocene.

*c. Dark-gray argillaceous beds with sandy intercalations.*

The next horizon consists of dark-gray to blueish-argillaceous beds to a thickness of 190—265 M. The demarcation with the underlying transitional gray *Globigerina* marls is not always distinct. The lithological character of the argillaceous beds however greatly differs from the white *Globigerina* marls.

*Molluscs* generally are very scarce, but in the section E—D an intercalation of a 5 M. thick sandy bed occurs out of which some *Molluscs* were collected.

| *Molluscs* of the Lower Pliocene beds E. of **Kedander** collected by Dr. J. W. M. NASH and determined by Dr. R. VON KOENIGSWALD. | Upper Miocene | Pliocene | Recent | Lower Pliocene Sangiran |
|---|---|---|---|---|
| Pleurotoma carinata Gray. . . . . | + | + | + | — |
| Marginella quinqueplicata Lam. . . | + | + | + | + |
| Hindsia tambacana Mart.. . . . . | + | + | — | + |
| Ranella subgranosa Beck . . . . . | + | + | + | + |
| „ nobilis Reeve . . . . . . | + | + | + | + |
| Casis pila Reeve . . . . . . . . . | + | + | + | + |
| Arca bistrigata Dkr. . . . . . . . | — | + | + | + |
| „ inaequivalvis Brug. . . . . . | — | + | + | — |
| Total 8 species det. . . . . . . . | 6 | 8 | 7 | 6 |

Not. det. Fusus sp. Dolium sp. Turritella sp. Tapes sp. Corbula sp., Corbicula sp., Tellina sp. and a coral: Flabellum sp.

The determined species being too few in number no certain conclusions are to be formed regarding the age of the beds.

The general habitus however in all features corresponds with that of the Lower Pliocene in the Mt. Pandan region. A Lower Pliocene age is most likely from the position of the beds underlying the next to be described marine *Mollusc*-bearing Pliocene beds.

Middle and Upper Pliocene and Pleistocene.

*d. Mollusc-bearing argillaceous tuffs, sands, conglomerates and breccias with limestone-intercalations and -concretions.*

The still insufficient data caused this group of Middle and Upper Pliocene and Pleistocene beds to remain unseparated on the map and in the sections. The total thickness in some cases even exceeds 1000 M.

Notwithstanding the difficulties met with in the field, to the opinion of the author it seems possible to state some welldefined horizons, that correspond with similar beds in other regions. Further investigations will certainly succeed in tracing the boundaries of these divisions on the map.

1. *Bedded limestone—Balanus limestone.* (*Base of Middle Pliocene*).

In the region N. of Kendeng mountains in the vicinity of Kembangan almost the basal bed of this group of layers is formed by *Balanus* limestone of 1½—2 M. thickness. A second bed of *Balanus* limestone of 2 M. thickness, occurs in a 100 M. higher stratigraphic horizon.

Both beds pass eastward into a shell bed or a shell-bearing limestone, but the lowest bed reappears as *Balanus* limestone in Banjuasin and Bakung.

The *Balanus* limestone was found in two spots near Pendem and Sukoredjo in the syncline separating the two anticlines, without sufficient dates however as to its position.

Two exposures of *Balanus* limestone were discovered in the axial beds of the southern anticline S. and E. of G. Putjangan which prevents ascertaining the stratigraphic position of this bed with regard to the base of the group of layers. They possibly correspond to the basal beds of Bedded limestone that could be traced over a great distance in the more Western region between G. Grobogan and G. Guwo along the southern slopes of this anticline.

Judging from the position with regard to the underlying Lower Pliocene argillaceous beds the Bedded limestone of the Southern region certainly corresponds with the lower bed of *Balanus* limestone of the Northern slopes of Kendeng mountains. Compared to the Bedded limestone and *Balanus* limestone of Kendeng mountains in the vicinity of Mt. Pandan it appears that they occupy about the identical stratigraphic horizon, with the exception of course of the second layer of *Balanus* limestone near Kambangan.

2. *Marine sands, argillaceous sands, and volcanic breccia (Middle and Upper Pliocene).*

A good general idea of the next group of layers in the Southern region is given by the following section.

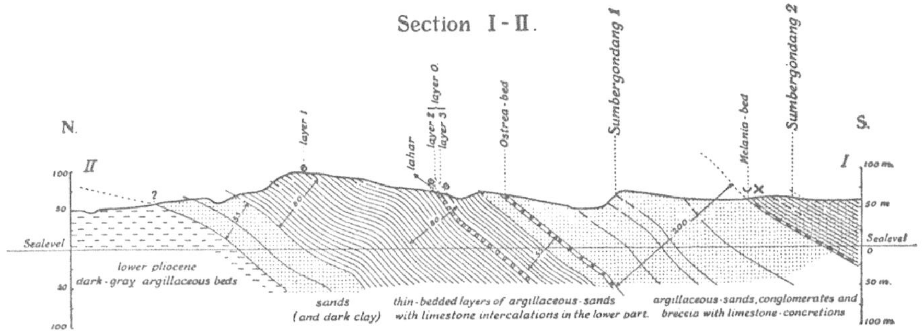


*Stratigraphical position of the fossilbearing layers of Sumberringin*

*Scale 1:10.000.*

*(Thickness of layers in meters)*

The transition of the Lower Pliocene argillaceous beds to the Middle Pliocene is not quite sharp. In the upper part of the argillaceous beds sandy intercalations appear, showing the beginning of changed conditions.

The bedded limestone occurring at the base of thinly bedded layers of argillaceous sands appears to consist of 4—6 separate layers to a thickness of 1—2 M. each. About 80 M. from the basal bed a *Mollusc*-horizon appears, hereafter denoted as **Sumberringin 1.** Two other *Mollusc*-bearing horizons occur in an 80 M. higher level and will be called **Sumberringin 2 and 3.** The lower of these two occurs just on top of a volcanic breccia („lahar") and consists of argillaceous coarse sandstone of 70—80 cM. thickness.

The superincumbent bed is a dark blue well-stratified clay of 4,5 M. thickness, covered by a 6—7 M. thick argillaceous sandstone which forms the third *Mollusc*-bearing horizon. The dip of the beds in 50° S. The overlying beds consist of sandstones with intercalations of clay.

The next covering beds are coarser, the lowest bed being a coarse-grained argillaceous breccia-sandstone dipping 52° S. E. and forming a ridge.

On top of it a fourth *Mollusc*-bearing horizon appears, a shell bed of 1,5 M. thickness containing *Ostrea lingua Sow.* This bed is still covered by 200 M. of argillaceous sands, conglomerates and breccia with limestone concretions before the first fresh-water horizon — consisting of a *Melania* bed — occurs.

The group of exclusively marine beds possesses a thickness of 415 M. starting from the basal beds of Bedded limestone. These basal beds have already been discussed as probably representing the horizon of Bedded limestone and the *Balanus* limestone in the vicinity of Mt. Pandan. The lowest fossil horizon therefore might correspond with the **Sonde** beds. When one considers the reduced thickness of these latter beds in Sonde and the Mt. Pandan area, there is some possibility of the higher horizons partly corresponding to the overlying **Butak** beds of volcanic character. This is supported by the intercalation of a volcanic breccia just below the second fossiliferous bed.

Corresponding evidence is given by the faunas of the mentioned fossiliferous horizons.

In accordance to the foregoing discussion, the horizons of *Molluscs* are treated separately, but the two faunas 2 and 3 are only $4\frac{1}{2}$ M. distant, taken in vertical sense and therefore will be combined.

It appears that the lower horizon of **Sumberringin** 1 contains 17 determined species of *Molluscs*, 59% of which are still living.

Only 7 species occur in **Sangiran,** 8 in **Bareng,** but as many as 11 species are known from **Sonde** or **Padasmalang.** This proves the facies to be identical to **Sonde** and confirms the allready supposed equal age. It is noteworthy that the three fresh-water *Molluscs*: *Paludina javanica v. d. Busch, Isidora sp.* and *Melania sp.* occur in the lake deposits of **Sangiran.** As to the species of

| Marine *Molluscs* from the Middle and Upper Pliocene of **Sumberringin** sampled by Dr. Ir. JAMES M.W. NASH and determined by Dr. R. VON KOENIGSWALD. | **Sumberringin** Layers 1 | 2 | 3 | 2+3 | Miocene | Pliocene | Recent | **Sangiran** Lower Pliocene | **Bareng** Middle Pliocene | **Sonde** Middle Pliocene |
|---|---|---|---|---|---|---|---|---|---|---|
| *Gasteropods* | | | | | | | | | | |
| Bulla ampulla Linn. | | + | + | | + | + | + | | | + |
| Plicarcularia globosus | | | + | | — | — | + | | | |
| Terebra cumingii Desh. | + | | | + | — | + | — | | | + |
| „ sp. | + | | | | | | | | | |
| Conus socialis Mart. | + | | | + | + | + | — | | + | + |
| „ ornatissimus Mart. | + | | | | + | + ? | — | | | |
| Surcula nodifera Lamk. | | | | + | + | + | + | | + | |
| „ javana Linn. | | + | | | — | + | + | | | |
| Pleurotoma gendinganensis Mart. | | | | + | — | + | — | + | + | + |
| Drillia losariensis Mart. | | | + | | — | + | — | | | |
| „ flavidula Lamk. | | | | + | — | + | + | + | + | + |
| Trigonostoma crispata Sow. | | | + | | — | + | + | | + | |
| Oliva funebralis Lamk. | | | | + | + | + | + | | | |
| „ ickei Mart. | | | | + | — | + | — | | | |
| Olivancillaria subulata Lamk. | | + | + | + | + | + | + | | + | |
| Marginella quinqueplicata Lamk. | | | | + | — | + | + | + | + | + |
| „ dactylus Lamk. | +[1] | | + | + | + | + | + | + | + | [2] |
| Voluta scapha Gmel. | | | | + | + | + | + | + | | [2] |
| Mitra circula Kiener. | | | + | | — | + | + | + | | + |
| „ sp. | | | + | | | | | | | |
| Melongena cochlidum Linn. | | | | + | + | + | + | + | + | |
| „ madjalenkensis Mart. | + | + | | + | — | + | — | | + | |
| Buccinum ventriosum Mart. | | | + | | + | + | + ? | | | |
| Dipsaccus canaliculatus Schum. | | | | + | + | + | + | | | |
| „ pankaensis Mart. | | + | + | | + | + | — | + | + | |
| Nassa Verbeeki Mart. | | + | + | + | + | + | — | + | + | + |
| „ gemmulata Lamk. | | | +[1] | +[1] | — | + | + | | + | |
| „ thersites Brug. | | + | | | — | + | + | | | + |
| Murex cf. junghuhni Mart. | | + | | + | + | — | — | | | |
| „ sp. | | + | | + | | | | | | |
| Eutritonium pileare Linn. | | | | + | + | + | + | + | + | |
| Persona reticulata Linn. | + | | | | + | + | + | + | | + |
| Hindsia gendinganensis Mart. | | | | + | — | + | — | + | | + |
| „ tambacana Mart. | + | | | | + | + | — | + | | + |
| Eutritofusus wanneri Fisch. | | | | + | — | + | — | | | |
| Ranella nobilis Reeve | | | | + | + | + | + | + | + | + |
| „ subgranosa Beck | + | | + | + | + | + | + | + | + | + |
| Cassis Pila Reeve. | +[1] | | + | | ? | + | + | + | + | + |
| Dolium costatum Desh. | + | + | | + | + | + | + | + | + | + |
| Cypraea sondeiana Mart. | | | | + | + | + | — | | + | + |
| Strombus isabella Lamk. | | + | + | + | — | + | + | | + | + |
| Potamides cheribonensis Mart. | | | | + | — | + | — | | + | |
| „ sulcatus Born. | | + | | | + | + | + | | + | |
| Telescopium titan Mart. | | +[1] | + | + | — | + | — | | + | |
| „ sp. | | | | + | | | | | | |
| Turritella vittulata Reeve | | | | + | — | + | + | | + | |

[1]) frequent.
[2]) from Padasmalang

| Marine *Molluscs* from the Middle and Upper Pliocene of **Sumberringin** sampled by Dr.Ir. JAMES M.W. NASH and determined by Dr. R. VON KOENIGSWALD | **Sumberringin** | | | | Miocene | Pliocene | Recent | **Sangiran** Lower Pliocene | **Bareng** Middle Pliocene | **Sonde** Middle Pliocene |
|---|---|---|---|---|---|---|---|---|---|---|
| | Layers | | | | | | | | | |
| | 1 | 2 | 3 | 2+3 | | | | | | |
| Solarium perspectivum Linn . . | | | | +[1] | + | + | + | + | + | + |
| Crucibulum extinctorium Lamk. . | | + | + | + | — | + | + | | + | |
| Crepidula walshii Herm. . . . . | | + | | + | — | + | + | | | |
| Xenophora calculifera Reeve. . . | + | | | + | — | + | + | | | + |
| Vanicoro sp. . . . . . . . . . . | | | | + | | | | | | |
| Natica melanostoma Gmel. . . . | | | | + | + | + | + | | + | + |
| „ powisiana Recl. . . . . . | | | | + | + | + | + | + | + | + |
| „ mamilla Linn. . . . . . . | | + | + | + | + | + | + | | + | + |
| „ ampla Phil. . . . . . . . | + | + | | | + | + | + | | | + |
| „ rufa Born. . . . . . . . | | + | | | + | + | + | + | + | + |
| Eulima sondeiana Mart. . . . . | + | | | | — | + | — | | + | + |
| Nerita signata . . . . . . . . . | | + | | | — | — | + | | | |
| „ chameleon Linn. . . . . . | | + | | | — | + | + | | | + |
| Trochus sp. . . . . . . . . . . | | + | | | | | | | | |
| Umbonium vestiarium Linn. . . | + | + | | | — | + | + | | | |
| Monodonta labio Linn. . . . . . | | + | | | — | + | + | | | |
| Ficus sp . . . . . . . . . . . | | + | | | | | | | | |
| *Lamellibranchiates* | | | | | | | | | | |
| Ostrea lingua Sow. . . . . . . . | | | | + | + | + | — | + | + | |
| „ promensis Noetling . . . | | | | + | + | + | — | + | | |
| Placuna placenta Linn. . . . . . | | | | + | + | + | + | + | + | |
| Plicatula imbricata Menke . . . | | | | + | — | + | + | | + | |
| „ granosa Mart. . . . . . | | | | + | + | — | — | | | |
| Amussium Hulshofi Mart. . . . . | | | | + | — | + | — | + | + | |
| Chlamys senatorius Gmel. . . . . | | | | + | + | + | + | + | + | + |
| „ kebolintangensis Mart. . | | | | + | — | + | — | + | + | + |
| Pinna vexillum Born. . . . . . . | | | | + | + | + | + | | + | |
| Arca oblonga Phil. . . . . . . . | | +[1] | + | + | + | + | + | | + | |
| Barbatia candita Helb. . . . . . | | + | | | — | — | + | | | |
| Limopsis venusta Mart. . . . . . | | + | + | + | ? | + | — | + | + | ? |
| Cardita boettgeri Mart. . . . . . | | | + | | — | + | — | | | |
| Crassatella radiata Sow. . . . . | | | | + | — | + | + | + | + | |
| Isocardia moltkiana Lamk. . . . | + | | | | + | + | + | | | |
| Gafrarium pectinatum Linn. . . . | + | | | | — | — | + | | | |
| Dosinia boettgeri Mart. . . . . . | | | | + | + | + | — | + | + | |
| Venus chlorotica Phil. . . . . . | | | | + | + | + | + | + | + | |
| Clementia papyracea Gray. . . . | | + | | | + | + | + | | | |
| Cyrena rustica Mart. . . . . . . | | | | + | + | + | — | | + | + |
| Corbula socialis Mart. . . . . . | + | | | + | + | + | — | + | + | |
| Total species 77 . . . . . . . . | | | | | 41+2 | 71 | 50 | 31 | 45 | 31+2 |
| 15 species not determined | | | | | | | | | | |
| *Brachiopods* | | | | | | | | | | |
| Terebratula sp. . . . . . . . . | | | + | | | | | | | |
| *Fresh-water Molluscs* | | | | | | | | | | |
| Isidora sp. 1 cf. Sangiran . . . . | + | | | | — | + | ? | | | |
| Melania sp. cf. Sangiran . . . . . | + | | | | — | + | — | | | |
| Paludina javana v. d. Busch. . . | + | | | | — | + | + | | | |

[1]) frequent

*Melania* it shows the same sharp keel and very much resembles the species that has been described from **Sangiran,** but of course owing to the great variability of *Melania* one is never sure that it really is the same species.

The fauna of **Sumberringin 2 and 3** yielded 71 species 66% of which are still living. This percentage is much higher than one of the already mentioned Pliocene horizons ever showed to possess. Only 29 species occur in the Lower Pliocene beds of **Sangiran,** 44 occur in **Bareng** and 31 in **Sonde** and **Padasmalang.**

Though the resemblance to **Bareng** is somewhat greater, the percentage of living species is so much more elevated that **Sumberringin 2 and 3** is to be considered as really representing a higher horizon of the Pliocene.

In a soon to be published report [1]) MARTIN described a similar fauna from **Kedungwaru,** a small campong occurring on a map published by COSIJN (1931) and representing a still farther Eastern part of Kendeng mountains N. of Modjokerto.

As the question is of great importance a discussion of the obtained results is given.

The list contains 40 determined and 9 unidentified species of *Gasteropods* and 10 determined and 13 unidentified species of *Lamellibranchiates.*

It appears that 32 species of *Gasteropods* are still living. Though this would give 80% living species MARTIN notwithstanding concluded to a Pliocene age as 28 species were already known from the Pliocene and 22 from the Miocene, whilst two species were exclusively known from the Pliocene.

In combining the *Gasteropods* with the *Lamellibranchiates* it will appear that of 50 determined species 35 or 70% are still living. As many as 8 recent species (of *Gasteropods*) were found to occur in this horizon.

Still accepting a Pliocene age this certainly proves the beds to belong to a very elevated horizon of the Pliocene.

The fossils have been collected on the road between Djetis and Sidotaka from a marine tuff sandstone, which according to the map and sections of COSIJN just underlies the main *Vertebrate* horizons of that region.

---

[1]) To be published in the „Jaarboek v.h. Mijnwezen in Ned. Ind."

This gives reason to the supposition that the *Mollusc* fauna of **Kedungwaru** corresponds with that of **Sumberringin 2 and 3** or even with a slightly higher horizon of the 255 M. thick series of marine beds that separate the latter fauna from the *Melania* bed in the Sumberringin region.

There is but one conclusion possible.

The series of beds to a total thickness of 415 M. between the basal bed of Bedded limestone and the fresh-water *Melania* bed embrace both Middle and Upper-Pliocene.

The lower *Mollusc* horizon of **Sumberringin 1** represents the **Sonde** beds, whereas the higher horizons of **Sumberringin 2 and 3** and probably **Kedungwaru** represent the marine facies of beds that in **Sangiran** and **Baringinan** are developed as lake deposits and in the **Mt. Pandan region** as volcanic breccias and tuffs.

North of Kendeng mountains the base of the Middle Pliocene beds near Simo is formed by *Balanus* limestone corresponding with the lowest of the beds occurring in Kambangan. In a stratigraphic level about 100 M. higher a shell bed occurs that takes the place of the second bed of *Balanus* limestone of Kambangan. This shell bed contains the fauna **Simo 1.**

In a 350 M. higher horizon another fauna **Simo 2 and 3** is found to occur in two separate layers to all probability intercalated in fresh-water beds.

Judging from the distance of the basal bed the fauna **Simo 1** almost corresponds with the fossiliferous horizon of **Sumberringin 1** and as one is to expect to find here the facies N. of Kendeng mountains a resemblance to the **Bareng** fossiliferous beds is likely. (See table page 119).

The list of marine Molluscs numbers 12 species, 7 or 58% of which are still living. This number of course is too small to deduce the age of the beds with sufficient certainty. The resemblance to the **Bareng** beds, however, is supported by 10 species or 83% being identical, whereas only 3 + 2 species or 42% are conformable to **Sonde** and only 2 species correspond to **Sumberringin 1**.

A Pliocene age is evident as all species are known from the Pliocene, whilst only 6 species likewise occur in Miocene beds and 5 species are extinct.

| Middle Pliocene marine mollusc-fauna **Simo 1**, collected by Dr. J. M. W. NASH and determined by Dr. R. VON KOENIGSWALD | Upper Miocene | Pliocene | Recent | **Bareng** | **Sonde** | **Sumber-ringin 1,** |
|---|---|---|---|---|---|---|
| Voluta scapha Gmel var. . . . . | + | + | — | | [1] | |
| Melongena madjalengkensis Mart | — | + | — | + | | + |
| Ranella subgranosa Beck. . . . | + | + | + | + | + | + |
| „ nobilis Reeve. . . . . . | + | + | + | + | + | |
| Dolium zonatum Green var. . . | — | + | + | + | | |
| Crucibulum extinctorium Lamk. | — | + | + | + | | |
| Placuna placenta Linn. . . . . | + | + | + | + | [2] | |
| Chlamys kebolintangensis Mart. | — | + | — | + | + | |
| Cardita boettgeri Mart. . . . . | — | + | — | | | |
| Crassatella radiata Sow. . . . . | — | + | + | + | | |
| Dosinia boettgeri Mart. . . . . | + | + | — | + | | |
| Venus chlorotica Phil. . . . . . | + | + | + | + | | |
| Total species 12 . . . . . . . | 6 | 12 | 7 | 10 | 5 | 2 |

Not determined: Anomya sp. Semifusus sp.

Fresh-water Molluscs. Paludina javanica v. d. Busch and Melania sp.

As a final conclusion it is most probable as well from the fauna as from the stratigraphic position, that the horizon of **Simo 1** corresponds with the Middle Pliocene **Bareng** beds.

The faunas of **Simo 2 and 3** will be jointly treated. They seem to occur in marine beds that alternate with fresh-water beds, that previously have been described by VERBEEK and FENNEMA (1896). As it is a fact that the beds occur in a much higher horizon than the first fauna, the evidence in the field shows them to be certainly younger than **Bareng.** (See table page 120).

The fauna of **Simo 2 and 3** much resembles the impoverished marine fauna that occurs intercalated in the lake deposits of **Sangiran** and **Baringinan.** Not only 8 species or 53,3% correspond but even two of the three prevailing species are the same. The list numbers 15 species, 9 of which or 60% are still living. Judging from the features of the fauna, it is of Pliocene age. As it is certain-

[1]) Padasmalang.

[2]) Sukun.

| Upper Pliocene marine *Molluscan* faunas **2** and **3** from **Simo** collected by Dr. J. M. W. NASH and determined by Dr. R. VON KOENIGSWALD | Upper miocene | Pliocene | Recent | **Baringinan** yellow arg. sand | **Sangiran** Yellow clay | **Baribis** *Turritella* bed |
|---|---|---|---|---|---|---|
| Marginella quinqueplicata Lam 2 . . . . . . . . . . . . | + | + | + | + | . | . |
| Cominella sangiranensis Mart. 2 [1]) 3. . . . . . . . . . . | — | + | — | + | + | |
| Melongena madjalengkensis Mart. 2. . . . . . . . . . . | + | + | — | + [1]) | + [1]) | |
| Potamides cf. Jenkinsi Mart. 2 | + | + | — | | | |
| „ cheribonensis Mart. 2. . . . . . . . . . . . . | — | + | — | | | |
| Turritella terebra Lamk. 2 . . | — | + | + | + | | + [1]) [2]) |
| Telescopium Titan Mart. 2 . . | + | + | + | | | |
| Placuna placenta Linn. 2 [1]) . . | + | + | + | + [1]) | + [1]) | + [1]) |
| Chlamys senatorius Gmel. 3. . | + | + | + | | | |
| Arca oblonga Phil. 2 3 [1]) . . . | + | + | + | + [1]) | + [1]) | + [1]) |
| „ inaequivalvis Brug. 2 . . | — | + | + | + | | |
| Tapes undulatus Born. 2 . . . | + | + | + | | | |
| Cyrena rustica Mart. 2 . . . . | + | + | + | | | |
| Corbula socialis Mart. 3. . . . | + | + | — | + | | |
| Pholas Hercules Mart. 2 . . . | — | + | — | | | |
| Total species 15 . . . . . | 10 | 15 | 9 | 8 | 4 | 2+1 |

Not determined Anomya sp. (2,3) Tellina sp. (3) Solen sp. (2).

ly younger than the Middle Pliocene **Bareng** horizon an Upper Pliocene age is probable. This would correspond with the age of the impoverished faunas of the lake deposits of **Sangiran-Baringinan** and of the *Turritella* bed of **Baribis.**

The result of the determination of the marine mollusc faunas supports the stratigraphical conclusions from the fieldwork and from the habitus of the beds. The marine lower part of the total division of 1000 M. thickness embraces both Middle Pliocene and Upper Pliocene beds. The combined thickness of Middle and Upper Pliocene beds S. of Kendeng mountains varies in the different sections between 415 and 575 M. N. of Kendeng mountains is sur-

[1]) Very frequent.
[2]) Turritella probably var. of terebra.

passes 450 M. It is to be noted that *Vertebrates* occur in the marine beds in the vicinity of Mt. Putjangan.

Though not regarding the discussed problems, the author wants to point out a singular fact concerning the occurrence of fossiliferous beds in the vicinity of **Simo.**

MARTIN (1919) published a list of molluscs collected by VAN DIJK at **Tambakbatu.** The list numbers 27 species 30% of which are still living, which gave reason to consider the fauna as Miocene.

In comparing this list with the two described faunas from **Simo,** it appeared, that only 2 species correspond with **Simo 1** and 2 other species are identical to **Simo 2 and 3.** As the river Tambakbatu was expressly explored in order to ascertain the horizon from which the collection was sampled by VAN DIJK, it was rather strange that the investigations failed to discover other than the two described Pliocene horizons. Either the horizon of Miocene molluscs was overlooked or the collection of VAN DIJK was derived from another unknown spot.

3. *Tuffs, sands, sandstones, argillaceous beds, breccias and conglomerates (Pleistocene including the Trinil beds).*

The upper part of the discussed division possesses a thickness of at least 500 M. on the southern slopes of Kendeng mountains. The material of these beds is generally even coarser than that of the underlying beds. The base of this group of beds is locally well characterized by the occurrence of a *Melania* bed, that could be traced from the vicinity of Sempal and Sumberringin to Munggang kerep.

*Vertebrates* abound in the sandy and argillaceous beds that overly the *Melania* bed.

From data collected by Dr. NASH it appears that the higher horizons contain marine *Molluscs*. This would prove *the existance of Pleistocene marine beds overlying the Trinil Vertebrate horizon.* It seems that the marine fossils are rather scarce and therefore no fauna is known yet. The author had no opportunity to investigate the spots where the marine *Molluscs* were reported to occur and therefore refrains from stating a definite opinion. If a fauna can be sampled of a sufficient number of species it will be of utmost interest for the determination of the age of the Trinil beds that now has to be deduced from other evidence.

The fauna of *Vertebrates* that occurs in the vertebrate-bearing horizons seems to be similar to that of Trinil. The specimens that were collected by COSIJN (1931) partly originate from the region N. of Djombang and were determined by UMBGROVE who stated the following species to occur:

Prionodon sp.
Crocodilus ossifragus Dubois.
Chitra sp.
Chitra cf. minor Jaekel.
Stegodon cf. Airawana. Mart.
Cervus cf. Lydekkeri Mart.
Bibos cf. palaeosondaicus Dubois.
Sus cf. macrognathus Dubois.
Buffelus sp.
Hippopotamus sp.

# CHAPTER VII

## SYNOPSIS OF THE STRATIGRAPHY

The stratigraphic table at the end of this chapter is a summary of the stratigraphy of the separate investigated regions. A mutual comparison gives the following results.

1. Upper Miocene, stage Tertiary f.

In Kendeng mountains from Surakarta to Djombang almost the same habitus of alternating tuffs and marls, volcanic breccias, sandstones and limestones prevails.

The presence of *Nephrolepidina sumatrensis Brady*, or of small species of *Lepidocyclina* and *Miogypsina* defines the age as Tertiary *e.* or *f*. But it is more likely that the beds belong to the younger one of these two stages as the superincumbent beds already belong to that part of the Upper Miocene, that does no more contain *Lepidocyclina* (Tertiary g). It is therefore probable that they correspond with the horizon represented by the **Tjilanang** mollusc-bearing beds in Western Java (32% living species). *Globigerines* and *Diatoms* occur.

2. Upper Miocene, stage Tertiary g.

In Kendeng mountains a division into several separate horizons is feasible.

*a. Gray Globigerina marls.*

This horizon is of great importance in the Northern anticlines. It is supposed to be contemporaneous to part of the white *Globigerina* marls in the Southern region. There is however a possibility that the upper part of the already described alternating tuffs and marls of the Southern anticlines may correspond to the gray *Globigerina* marls.

*b. White Globigerina marls.*

This horizon is widely distributed in Kendeng mountains. *Globi-*

*gerines* prevail, but many other small *Foraminifera* are also present. KOCH determined 107 species probably derived from this horizon and by comparison with other faunas of Pliocene age, concluded the beds to be either Miocene or Lower Pliocene. It must be admitted however that there is no absolute certainty as to the exact horizon from which these *Foraminifera* have been collected.

*c. Globigerina limestone.*

In some regions the *Globigerina* limestone tops the white *Globigerina* marls, in others it fails to occur. This horizon by the absence of other deposited matter, indicates the zenith of a period of less volcanic activity, already begun in the time when the white *Globigerina* marls were deposited.

*d. Coral limestone.*

The *Coral* limestone either occurs at the top of the *Globigerina* limestone or replaces it. It is, with a few exceptions, restricted to the Southern anticlines, but even there is absent in the most Western and farthest Eastern part of the investigated regions of Kendeng mountains.

*e. Transition marls.*

The uppermost horizon of gray sandy marls still contains *Globigerines*. A decrease in number and size however indicates the end of a period in which *Globigerines* prevailed. The sandy character of the beds is brought about by an increase of volcanic matter, which proves a revival of volcanism.

The wholly changed character of the superincumbent beds — the reason why the discussed horizon was called transitional — in the opinion of the author is to be explained by the totally different conditions of sedimentation. In the preceding period an open ocean existed. The rising movements in the Southern part of Java reduced the sea to a narrow arm in Northern Java.

The upper Miocene age of the enumerated beds is to be concluded from the following plea.

The underlying beds, by the presence of *Lepidocyclina* are proved to contain no younger horizons than the Tertiary *f.*, of which the **Tjilanang** beds (32% living species) are a representant).

On the other hand the — next to be mentioned — superincumbent beds are to be described as Lower Pliocene as follows from the presence of the molluscan fauna of **Sangiran** with 45% living species.

The discussed *Globigerina*-bearing horizons (including the *Coral* limestone) just fill the gap between the *Lepidocyclina*-bearing part of the Miocene and the Lower Pliocene. They must correspond with that part of the Miocene in which the *Mollusc*-faunas of **Tjiodeng** 42% living species) and **Palabuanratu** (38% living species) occur.

There might of course remain some doubt as to the exact demarcation between the Upper Miocene and the Lower Pliocene. Considering the habitus of the Transition beds, the still predominant presence of *Globigerines*, which certainly is the most characterizing feature of the Upper Miocene (Tertiary stage *g*), marks this horizon as still belonging to the Upper Miocene.

3. Lower Pliocene.

The Lower Pliocene generally appears as dark gray argillaceous beds. Of the three described fossiliferous horizons — including **Kaliuter** and **Kedander** — only the *Turritella* bed of **Sangiran** (45% living species) includes sufficient determined *Molluscs* to regard the percentage of living species as rather safe.

The established figure fits in between the Upper Miocene faunas of **Tjiodeng** (42%) and **Palabuanratu** (38%) and the extremely safe Pliocene horizon of **Sonde** (53%).

Though Martin (1919) mentions an equal percentage for **Parungponteng** (45%) it appears that this Miocene horizon rather corresponds with the likewise Miocene fauna of **Tjadasngampar** with only 27% living species. Therefore the percentage of **Parungponteng** needs some caution as to its reliability.

It has been shown by Martin (1919) that **Tjiodeng, Palabuanratu** and **Parungponteng** possess more affinity to acknowledged Miocene faunas, whereas on the other hand, the author proved the fauna of **Sangiran** to be more related to Pliocene faunas, by the higher ratio of Pliocene species with regard to Miocene species.

Moreover, since Fischer (1921) mentions 46,8% as the number of living species in the Pliocene Fufa beds, there is no reason to regard the percentage found in the *Turritella* bed of **Sangiran** as rather too low to represent Pliocene beds. Tesch (1920) likewise described a Pliocene fauna from Timor not containing more than 43% living species.

The lower percentage of living species and the stratigraphic position of the dark argillaceous beds — underlying the fossiliferous horizon of **Bareng** — marks them as older than **Sonde,** so there seems to remain no doubt that they represent an older horizon of the Pliocene.

4. Middle Pliocene.

*a. Conglomeratic limestone, Bedded limestone and Balanus limestone.*

In the Northern Kendeng mountains N. of Mt. Pandan Bedded limestone locallly occurs, overlying the dark argillaceous beds. N. of Djombang *Balanus* limestone, locally passing into a shellbed, occupies a similar position, Near Kambangan however a second bed of *Balanus* limestone appears to occur in a somewhat higher level.

In the Southern Kendeng mountains Conglomeratic limestone containing pebbles of Miocene marls often forms the base of the transgressing Middle Pliocene beds. This transgression indicates the presence of a local unconformity caused by an earlier upheaval (probably at the end of the Miocene) of part of the Southern Kendeng mountains.

The basal Conglomeratic limestone is sometimes directly covered by Bedded limestone or by *Balanus* limestone.

In N. Surakarta, where the *Balanus* limestone is well developed and forms a very persisting horizon, it bears a conglomeratic habitus and occurs as the basal bed of the Middle Pliocene.

N. of Djombang, where no hiatus in the succession of the beds seems to exist, either Bedded limestone or *Balanus* limestone appears to overlie the dark argillaceous beds of the Lower Pliocene.

*b. Argillaceous and sandy beds.*

Argillaceous and sandy beds form the main fossilliferous horizon of the Middle Pliocene. The wellknown mollusc-bed of **Sonde** (53% living species) is to be regarded as a reliable horizon. **Waled** (54%) and **Pangka** (61%) are of corresponding facies and age.

**Bareng** (53%) represents another facies of the same horizon and rather resembles the faunas of **Tjidjurai** (51%), **Tjidjadjar** (56%) and **Gombel** (51%).

**Bumiaju** (56%) resembles **Tjidjurai, Simo 1** (58%) more or less corresponds to **Bareng,** whereas **Sumberringin 1** (59%) and the *Coral* reef of **Baribis** (56%) show more resemblance to the facies of **Sonde** and **Waled.**

The equal number of living species found in **Bareng** and **Sonde** leaves no doubt as to both faunas, though differing in species, representing the same horizon of the Pliocene, as the number of species in both cases is more than a hundred and therefore the percentage of living species is to be regarded as rather safe.

After the publications of UMBGROVE (1929) and MARTIN (1931) no doubt remains as to the Pliocene age of the **Sonde** fauna.

5. Upper Pliocene.

The Upper Pliocene sediments show a great variety. In the opinion of the author the orogenic movements, that caused the folding of the Tertiary beds were strongly active during the Upper Pliocene.

The Pleistocene Trinil beds are characterized as being mainly *denudation* products on account of their content of *detritus* from Miocene and Pliocene beds, mixed with andesitic products *washed down* from the slopes of volcanoes. In several spots they distinctly overlie unconformably the older beds, for instance N. of Kendeng Mountains and in the Trinil region. The unconformity in **Trinil** where the Pleistocene Trinil beds directly overlie the Middle Pliocene Sonde beds indicates that the folding movements acted in the intervening period. It has been proved now that the existing stratigraphic gap in Trinil has to be bridged by an interjacent series of Upper Pliocene layers, that now have been described from several localities. Their occurrence is restricted to synclinal regions. It is hardly conceivable that folding should have taken place in a relatively short period e.g. in the short time between the forming of the Upper Pliocene beds and the Pleistocene Trinil beds. Moreover it is distinctly shown in Sangiran that the Upper Pliocene lake deposits without a break gradually pass into the Pleistocene Trinil beds.

The simultaneous activity of orogenic movements explains why the Upper Pliocene beds possess an enormous thickness in some synclines and on the other hand show a restricted horizontal extension. Meanwhile it is now clear why the habitus is so much different in the various regions where they occur. The proved considerable increase of volcanic activity in the Upper Pliocene supports the supposed simultaneousness of orogenic movements.

In **Bumiaju** a mighty series of terrestrial and lacustrine beds shows to have been deposited in a sinking synclinal region. At the base it starts with vertebrate-bearing conglomerates and sandstones containing a fresh-water fauna with 69% living species. This horizon is covered by tuffs and conglomerates.

Northern Surakarta appears to have been covered by a big lake, the deposits of which possess a thickness of over 200 M. In **Sangiran** (fresh-water *Molluscs* with 36½% living species) and **Baringinan** (fresh-water *Molluscs*).

In **Kaliuter** the supposed Upper Pliocene beds consist of freshwater sandstones and tuffs. Neither has the age of the beds been proved in **Mt. Patih-ajam,** where basal beds of argillaceous lacustrine deposits somewhat resemble the Upper Pliocene beds of **Bumiaju** and the higher horizons of volcanic breccia were supposed to correspond with the breccias of the **Butak**-beds in the vicinity of Mt. Pandan.

In S. Kendeng mountains in the vicinity of Mt. Pandan the lowest horizon of the so called **Butak** beds contains *Melania* and *Corbicula* beds. Volcanism was very active in this region. The bulk of the **Butak** beds is formed by volcanic breccias and tuffs sometimes alternating with conglomerates. Beds of volcanic breccia of minor importance occur in **Sangiran, Baribis** and in **Kendeng mountains N. of Djombang.**

The described terrestrial and lacustrine habitus of the beds does not include that marine deposits were absent. An *invasion of the sea* in the lake deposits of **Sangiran** and **Baringinan** proves that *other regions were still covered by the sea.* The marine bed contains an *impoverished marine fauna* that greatly resembles a corresponding fauna in **Baribis** (*Turritella* bed) and in **Simo 2, 3.** In these last mentioned spots the marine deposits were found to alternate with fresh-water conglomerates.

Similar conditions existed in the upper horizons of **Tjidjurai.** And even in the region of **S. Kendeng mountains** where the main deposits already have been described as terrestrial volcanic products and lacustrine beds, the neighbourhood of the sea was proved in the syncline of Soko, where the lowest bed of volcanic breccia was found to be intercalated in marine deposits.

Marine deposits were still being formed in the broad synclinal region between Tjepu and Bodjonegoro N. of Kendeng mountains. They sometimes contain much volcanic matter e.g. in the vicinity of Bara.

Marine deposits N. of Djombang contain a marine fauna in **Sumberringin 2, 3** that possesses 66% living species of of *Molluscs*. Somewhat farther East a corresponding fauna of **Kedungwaru** was found to contain as many as 70% living species.

Though the next to be described **Trinil** beds appear to have been still influenced by folding movements following the big folding of the Tertiary beds the fact that they unconformably overlie older beds in several regions proves the maximum of orogenic movements to be posterior to the forming of the Trinilbeds.

The described Upper Pliocene beds were certainly much stronger folded.

The alternating marine and terrestrial deposits of **Baribis** show vertical dips. The upper part of the **Tjidjurai** beds dips conformably to the lower horizons of Middle Pliocene beds.

The terrestrial and lacustrine deposits of **Bumiaju** are strongly folded. The lakedeposits of **Sangiran** occur in the slopes of a domeshaped anticline. But what is more convincing yet, the corresponding lake deposits of **Baringinan** are unconformably covered by the **Trinil** beds. The volcanic breccias and tuffs of the **Butak** beds, in the vicinity of Mt. Pandan show a dip of 15° near Mt. Butak whilst the **Trinil** beds in Kedungbrubus dip only 6°.

That the Butak beds do not belong to the series of Trinil beds though an unconformity is absent and they likewise contain *Vertebrates*, is shown by the fact that in the syncline of Soko the lowest bed of breccia is intercalated between the Pliocene marine beds.

Altogether the different beds that have been compiled in this division do still belong to the rather strongly folded series whereas the superincumbent Trinil beds were in several places unconformably deposited upon already folded beds, though they seem to have been still influenced by posterior slight folding movements.

The supposition of VERBEEK (1896—1908) that a demarcation between Tertiary and Quaternary beds could follow from the dip of the beds is erroneous as to the assumption of a horizontal position of the Quaternary beds. The problem is one of relativity. If real strong dips occur a Quaternary age of the beds is rather unlikely, but it is conceivable that Quaternary beds show dips even up to 30°.

The percentage of living species of Molluscs in the **Sonde** beds

(53%) and corresponding beds (varying from 51 to 61%) shows that a great stratigraphic gap exists between this aknowledged Pliocene horizon and the Holocene (Upper Quaternary) beds of **Batavia** (87%) and **Grissee** (90%) the age of which is esteemed at about 4000—6000 years ago, according to the theory of DALY (1920).

It was certainly not likely that this gap should merely embrace the Pleistocene. It was rather more probable that part of this gap is to be filled by the Upper Pliocene. It is now proved that fauna's exist of **Sumberringin 2, 3** (66%) and **Kedungwaru** (70%) of Upper Pliocene age previous to the forming of the Trinilbeds.

6. Pleistocene.

*a. Riverconglomerates, sands, tuffs, volcanic breccias (Trinil beds).*

The Pleistocene vertebrate-bearing beds in several spots unconformably overlie the older horizons. Of course an unconformity does not exist in regions where the sedimentation was uninterrupted. Such is the case in volcanic regions, for instance S. Kendeng mountains in the vicinity of Mt. Pandan and in Mt. Patih-ajam. Neither was an unconformity probable in Kendeng mountains N. of Djombang and in originally synclinal regions as in Sangiran.

But the unconformity was very distinct in Baringinan, in the region between Gesi and Ngawi — including the famous spot of Trinil — and E. of Ngawi.

The unconformity is even very strong N. of Kendeng mountains, where vertically dipping Miocene beds (Randublatung) are covered by only slightly dipping Trinilbeds, though of course in synclinal regions as N. of Bareng-Tondomuljo, the difference in dip is much less.

Confirming evidence is brought by the habitus of conglomeratic and sandy beds. These layers in several instances were found to contain detritus of Miocene beds and Miocene and Pliocene marine fossils.

In Patih-ajam and Sangiran the habitus of the beds — river deposits containing matter of mainly volcanic origin — much resembles that of Southern Kendeng mountains.

N. of Kendeng mountains, however, though still volcanic matter is present to a considerable content, the habitus is different as much more detrital sediments occur. In the region of Tjepu the beds are even exclusively composed of detrital products of Miocene and Pliocene beds.

In the region N. of Djombang marine *Molluscs* and *Globigerines* occurring in beds overlying the main vertebrate horizons may point to a partly marine facies of the Trinil beds.

*b. Tuffs, volcanic breccias, sandstones.*

In Mt. Patih-ajam, Sangiran and in the vicinity of Mt. Pandan the vertebrate-bearing Trinilbeds are covered by a series of tuffs and sandstones sometimes with intercalations of volcanic boulderbreccia, that still show a very slight dip sometimes even reverse to the original natural dip of the beds. They are void of vertebrate-remains and were called Notopuro beds in the Mt. Pandan region, where they seem to be of equal age as the last products of Mt. Pandan volcano, consisting of breccia and tuffs.

7. Holocene.

The marine deposits in the subsoil of **Batavia** to a depth of less than 6 M. and of **Grissee,** according to the theory of Daly, that has been confirmed by different facts in Java, date from a period no more distant than 4000—6000 years ago.

8. The vertical distribution of vertebrates:

*a. Middle Pliocene.*

There is no sure evidence of *land-Vertebrates* occurring in Java during the Middle Pliocene.

*Vertebrates* (*Stegodon*) indeed have been found in the alternating marine and fresh-water beds of **Tjidurai.** The determination of the marine *Molluscs* (51% living species) indicated a Middle Pliocene age of the marine beds. But it has to be pointed out, that the greater part of the *Molluscs* occurs in the *deeper horizons* that *underlie* the *vertebrate*-bearing fresh-water beds and that only a very few species of marine *Molluscs* originate from the alternating upper layers. It is conceivable that the lower horizons represent the Middle Pliocene, whereas the upper layers are of Upper Pliocene age.

A femur of *Rhinoceros* was discovered by the author in the **Ba -**

**reng** beds near Bara but this also does not prove a Middle Pliocene age of the *Vertebrates* in Java as this femur possibly is only part of a floating carcass that may originate from Borneo or another part of the Archipelago that already was connected with the Asiatic continent.

*b. Upper Pliocene.*

There is certainly much more evidence with regard to *Vertebrates* occurring in the Upper Pliocene beds of Java.

The Upper Pliocene terrestrial conglomerates that alternate with marine beds in **Baribis** contain molars of *Stegodon* and *Cervus.*

The terrestrial conglomerates of **Bumiaju** show a rich fauna including *Mastodon perimensis*, *Stegodon airawana Mart.* and *Hippopotamus.* In the lake deposits of **Sangiran** remains of *Crocodilus* and *Cervus* have been discovered.

The volcanic **Butak** beds contain *Stegodon* and *Crocodilus* near Mt. Butak.Several *Vertebrate*-remains were found in the marine beds of **G. Putjangan.**

The invasion of a *Vertebrate* fauna in Java that originated from the Asiatic continent, without any doubt occurred in a period *anterior* to the forming of the **Trinil** beds. It dates as far back as the Upper Pliocene.

The intercalation of marine beds between the deltoid deposits of **Baribis** and **Tjidjurai,** the invasion of the sea in the lake deposits of **Sangiran** and **Baringinan,** and the marine habitus of the Upper Pliocene beds **N. of Kendeng mountains** and **N. of Djombang** indicate that the described Pliocene sea arm in the Northern part of Java is to be supposed as still to have existed during the Upper Pliocene.

There is reason to accept the invasion to have happened via Western Java. There is a possibility of a landbridge, that connected Java with Sumatra, to have existed even anterior to the Upper Pliocene as the Pliocene faunas of Southern Bantam show very small resemblance to that of **Tjidjurai** as already has been pointed out by MARTIN (1926).

Of the Molluscs occurring in **Tjidjurai** only $\frac{1}{8}$ correspond with **Tjikeusik** (Bantam) and only $^1/_7$ with **Tjimantjeurih** (Bantam). There is no evidence where another landbridge was formed at the commencement of the Upper Pliocene, that connected West-

Additional material from *The Age of Pithecanthropus*,
ISBN 978-94-017-5674-7 (978-94-017-5674-7_OSFO1),
is available at http://extras.springer.com

EXTRA
MATERIALS
extras.springer.com

ern Java (eventually via Sumatra) with the Asiatic continent, nor is there any proof as to the origin of this landbridge. But the author is inclined to accept orogenic movements during the Upper Pliocene having caused an extension of the land in Java and Sumatra, as certainly in Java the regression of the sea from anticlinal and some synclinal regions during the Upper Pliocene is very marked. On the other hand the simultaneously increased volcanic activity may have favoured the filling up of the sea arm that separated Java (or Sumatra) from the Asiatic continent.

The repeated regressions of the sea during the Pleistocene glaciations that have been brought forward by MOLENGRAAFF (1923) may have had some influence upon the Vertebrate fauna by enlarging the roads of communication with the Asiatic Continent, still the centre of interest lies with the statement of an Upper Pliocene origin of the *Vertebrate* population of Java.

*c. Pleistocene.*

The *Vertebrate* beds of **Mt. Patih-ajam** and the main vertebrate bed in **Sangiran** and **Baringinan** are to be considered as of the same age as the **Trinil** beds South of Kendeng mountains from Gesi to the vicinity of **Trinil.** North of Kendeng mountains the occurrence of Vertebrates has been described from **Ngembak hill, Randublatung, Tinggang, Tjepu** and **Bareng.**

The **Trinil** beds N. of Djombang are perhaps partly of marine origin. The occurrence of Vertebrates is abundant in the lower part of the Pleistocene especially in the river deposits.

Volcanic tuffs and breccias prevail in the higher horizons were *Vertebrate* remains are seldom met with.

The occurrence of *Vertebrate* remains on the banks of **Tjitarum** river mentioned by STEHN and UMBGROVE (1929) is to be considered as *younger* than the **Trinil** beds by the absence of *Stegodon, Elephas* and *Hippopotamus.*

# CHAPTER VIII

## THE AGE OF PITHECANTHROPUS ERECTUS

The age of the **Trinil** beds, the discussion of which has been held back to this final chapter, is the main problem that wants to be solved, as it meanwhile involves the question of the age of *Pithecanthropus erectus*.

It has already been mentioned, that between the **Trinil** beds — in which Pithecanthropus erectus was discovered — and the in Trinil almost directly underlying **Sonde** beds, a great stratigraphic gap exists, that has to be bridged by a separate group of layers.

This group of layers is represented in **Bumiaju** by a mightly complex of terrestrial and lacustrine alternating layers of conglomerates, argillaceous beds and tuffs that contain *Vertebrates*. In **Surakarta** an interesting succession of lake deposits occurs, in which a *marine invasion* of short duration has been proved to exist. A third facies was found to exist in the vicinity of Mt. Pandan, where the identical horizon consists of volcanic breccias and tuffs (**Butak** beds). A fourth variety of facies is developed **N. of Djombang** where marine deposits occur that appeared to contain 66% living species of *Molluscs* in the **Sumberringin horizons** 2 and 3 and 70% recent forms in the fauna of **Kedungwaru.**

The existance of these marine faunas with 66% respectively 70% of living species contemporaneous to horizons, *older* than the **Trinil** beds, indicates, that *if ever a marine fauna of the age of the Trinil beds is discovered in Java, it will show to contain more than 70% living species of Molluscs.*

Up till now no marine fauna has been collected of the age of the Trinil beds. There is however some possibility of discovering it in the region N. of Djombang where *marine* beds have been stated to occur covering the *Vertebrate* horizon of the **Trinil** beds.

The conclusion that a still to be discovered marine fauna corresponding with the Trinil beds necessarily must contain more than the 70% of living species that occur in the Upper Pliocene of **Kedungwaru** pleads for a Pleistocene age of the Trinil horizon, as this to be expected very high percentage fills the gap between the Upper Pliocene and the post Pleistocene beds of **Grissee** (90%) and **Batavia** (87%).

In a former chapter the contrary opinion of HILBER has been discussed, who doubted the conclusions of MARTIN–ICKE with regard to the Pleistocene age of the Trinil beds that had been based upon the percentage of living species of fresh-water gasteropods (87½%) occurring in those beds. He held forth the uncertainty of the figures deducted from the fresh-water Molluscs in stating an example from Italy.

It is now evident, that the probable percentage of living species of *marine Molluscs* that is to be expected in marine beds of the age of the **Trinil** beds lies between 70% and 86% and certainly will not differ much from the percentage of 83% of *fresh-water Molluscs* that was stated by HILBER for the Trinil beds.

A similar conclusion can be deduced from a comparison of the fresh-water *Molluscs* of the **Trinil** beds and those of the Upper Pliocene lake deposits of **Sangiran** and **Baringinan.**

Of the compared species only 4 species are proved to be corresponding. Even if the undetermined species of *Limneus, Planorbis, Isidora* and *Unio* were found to be similar, this would not give more than 8 species of the **Trinil** beds corresponding to those of the lake deposits of **Sangiran** and **Baringinan.** (See table page 136).

On the other hand the determined species of *Melania* of **Sangiran** — with the exception of two widely distributed forms — do not occur in the **Trinil** beds. The absence of these species marks the **Trinil** beds as *much younger* than the lake deposits of **Sangiran.**

In the lake deposits of **Sangiran** and **Baringinan** 6 fully determined species out of a total of 16 occurring species are extinct. Whatever the result of the determination of the remaining forms may be, it proves that the percentage of living species can never exceed 62½%.

The existance of an Upper Pliocene fresh-water fauna in Java showing a percentage of living species of 62½% or less, proves the age of the **Trinil** beds, in which 83% of living species of fresh-water *Molluscs* is to be considered as a minimum, to be Pleistocene.

Fresh-water *molluscs* of the **Trinil** beds after MARTIN-ICKE and MARTIN:

| | Miocene | Pliocene | Pleistocene | Recent |
|---|---|---|---|---|
| **Trinil :** | | | | |
| Helix rotatoria v. d. Busch | — | — | + | + |
| Limneus rubiginosa Michel | — | — | + | + |
| Planorbis tondanensis Quoy et Gaim | — | — | + | + |
| Physa sp. (= Isidora sp.) | — | + | + | ? |
| Melania testudinaria v. d. Busch | — | + | + | + |
| Melania tuberculata Müller | + | + | + | + |
| „ granum v. d. Busch | — | + | + | + |
| „ savinieri Brot | — | — | + | + |
| „ verrucosa Hinds | — | — | + | + |
| „ sp. | ? | ? | + | ? |
| Bithynia truncata Eyd et Sow. | — | — | + | + |
| Paludina javanica v. d. Busch | — | + | + | + |
| Ampullaria scutata Mouss. | — | — | + | + |
| Unio productus Linn. | — | — | + | + |
| „ trinilensis? Dubois | — | — | + | ? |
| **South of Mt. Pandan** | | | | |
| Bulimus citrinus Brug. | — | — | + | + |
| Melania infracostata Mouss. | — | — | + | + |
| Ampullaria ampullacea Linn. | — | — | + | + |
| **Ngembak hill** | | | | |
| Corbicula exporrecta Mart. | — | + | + | — |
| | 1 | 6+1 | 19 | 15+3 |

Freshwater *molluscs* of the lake deposits in **Sangiran-Baringinan** after MARTIN added by the author:

| | Miocene | Pliocene | Pleistocene | Recent |
|---|---|---|---|---|
| **Sangiran :** | | | | |
| Isidora sp. | — | + | + | ? |
| Melania testudinaria v. d. Busch | — | + | + | + |
| soloënsis Mart. | — | + | — | — |
| tjemoroënsis Mart. | — | + | — | — |
| bojolaliensis Mart. | — | + | — | — |
| fennemai Mart. | — | + | — | — |
| granum v. d. Busch | + | + | + | + |
| kritjianensis Mart. | — | + | — | — |
| nov. sp. | + | + | ? | ? |
| Paludina javanica v. d. Busch | — | + | + | + |
| Corbicula exporrecta Mart. | — | + | + | — |
| Unio sp. | — | + | ? | ? |
| **Baringinan** | | | | |
| Limneus sp. | — | + | ? | ? |
| Planorbis sp. | — | + | ? | ? |
| Segmentina sp. | — | + | ? | ? |
| Hydrobia sp. | — | + | ? | ? |
| | 1 | 16 | 5+6 | 3+7 |

The author therefore *strongly supports* the conclusions of MARTIN–ICKE regarding a Pleistocene age of the **Trinil** beds. Both arguments, as well the first, that proved *a percentage of living species superior to 70% to be expected in a marine fauna of the age of the* **Trinil** *beds,* as well as the second, referring to *a percentage of 83% of living species in the fresh-water fauna of* **Trinil** *as compared to only 62½% or less in the Upper Pliocene beds of* **Sangiran** *and* **Baringinan,** *show the Pleistocene age of the* **Trinil** *beds.*

This conclusion is not contradicted by the results of the investigations regarding the *Vertebrates.* According to competent authors (STREMME) the features of the **Trinil** fauna show *more affinity* to those of the Pleistocene **Narbada** fauna than to those of older Asiatic faunas.

The stratigraphic position of the **Trinil** beds underlying a mighty complex of volcanic sediments (**Notopuro** beds) in the vicinity of Mt. Pandan and in the basal part of an at least 500 M. thick complex of layers in the region N. of Djombang, shows them to occupy a rather low horizon of the Pleistocene. This meanwhile is also evident from the accompanying fauna (*Mastodon? sp. Stegodon airawana* and *trigonocephalus* and *Hippopotamus sp.*)

As to the sometimes expressed possibility of fossil vertebrates of older beds to have been washed into the Trinil beds, the author wants to points out that the often wellpreserved condition of skullparts, molars, and tusks of *Proboscideans* and remains of other *Vertebrates* excludes an older origin of these fossils occurring in the Trinil beds. A fossil tusk of some M. length can only be transported with the utmost care and generally appears to have already been broken into adjoining parts in the spot where it occurs, how can it have been washed down in its entire length from older beds in a fossil state?

The Upper Pliocene origin of the first invasion of vertebrates in Java makes it rather improbable that a separate evolution of the *Vertebrate* fauna should have happened, as the discussion of the influence of iceperiods upon sealevel variations shows, that in the ensuing periods the communication with the Asiatic continent was rather better during the periods of glaciation. If really adequate proofs can be brought forward for insular nanism in the sense as has been used by DIETRICH (1924) this evolution rather occurred in the period between the Upper Pliocene invasion and the Lower Pleistocene general expansion of the *Vertebrate* fauna in Java.

The foregoing arguments show *Pithecanthropus erectus* to have lived during the Lower part of the Pleistocene accompanied by the fauna that has been described from the **Trinil** beds.

The discussed more primitive features of the skullcap of *Pithecanthropus erectus* gave rise to the possibility of its representing an earlier hominid form than *Sinanthropus Pekingensis* or *Eoanthropus dawsoni*.

The presence of *Mastodon? Stegodon* and *Hippopotamus* in the fauna of the **Trinil** beds and the undoubtedly low stratigraphic position of this horizon in the Lower Pleistocene, confirm the statement that *Pithecanthropus erectus Dubois* is still to be considered as the oldest hominid form discovered.

# LITERATURE

1857 F. JUNGHUHN. Over de fossiele zoogdieren te Patikajam. Nat. Tijdschr. v. Ned. Ind. XIV, p. 215.

1858 A. Baron SLOET V. OLDRUITENBORGH. Mededeeling over fossiele werveldierresten. Nat. Tijdschr. v. Ned.-Ind. XVI, p. 70.

1864 P. E. C. SCHMÜLLING. Mededeeling over fossiele werveldierresten. Nat. Tijdschr. v. Ned. Ind. XXVII, p. 399.

1867 RADEN SALEH. Over fossiele zoogdierbeenderen in het regentschap Sentolo Jogjakarta. Nat. Tijdschr. v. Ned. Ind. XXIX, p. 423, 428, 434—435.

1867 RADEN SALEH. Over .............. fossiele beenderen van den Pandan. Nat. Tijdschr. v. Ned. Ind. XXIX, p. 450—451.

1867 P. VAN DIJK. Geologische beschrijving der Res. Djokjakarta. Tijdschrift v. Nijverh. en Landbouw in Ned. Ind. XIV, p. 192. The same in Jb. Mijnw. 1872, I p. 183.

1875 J. CROLL. Climate and time. London.

1882 A. PENCK. Schwankungen des Meeresspiegels. Jahrb. d. Geogr. Ges. zu München VII, p. 47.

1883—1887. K. MARTIN. Palaeontologische Ergebnisse von Tiefbohrungen auf Java. Samml. Leiden Ie Serie, Bd. III.

1884 K. MARTIN. Ueberreste vorweltliche Proboscidier von Java und Banka. Samml. Leiden Ie Ser. Bnd. IV, p. 1—24.

1886 K. MARTIN. Fossiele Säugetiere von Java und Japan. Samml. Leiden Ie Ser. Bd. IV, p. 25—99.

1888 K. MARTIN. Neue Wirbelthierreste von Patihajam auf Java. Samml. Leiden Ie Ser., Bd. IV, p. 87—116.

1889 E. DUBOIS. Over de wenschelijkheid van een onderzoek naar de diluviale fauna van Ned. Indië. Nat. Tijdschr. v. Nederl. Ind. XLVIII, p. 148—165.

1890 K. MARTIN. Ueber neue Stegodon-Reste aus Java. Verh. Kon. Akad. v. Wetensch. Amsterdam. Versl. Afd. Natuurk. XXVIII

1892 E. DUBOIS. Voorloopig bericht omtrent het onderzoek naar de pleistocene en tertiaire Vertebratenfauna van Sumatra en Java gedurende het jaar 1890. Natuurk. Tijdschr. v. Ned. Ind. LI. 1892, p. 93—100.

1894 E. DUBOIS. Pithecanthropus erectus, eine menschenähnliche Uebergangsform aus Java. Batavia. Also in Jb. Mijnw. 1895 Wet. Ged., p. 5—77.

1896 R. D. M. VERBEEK en R. FENNEMA. Geologie van Java en Madoera. Amsterdam.

1900 K. MARTIN. Die Eintheilung der versteinerungsführende Sedimente auf Java. Samml. Leiden 1ste Ser. Bd. VI., p. 135—245.

1904 F. FRECH. Lethaea geognostica III. Erste Abt., p. 30.

1907 E. DUBOIS. Eenige van Nederlandschen kant verkregen uitkomsten met betrekking tot de kennis der Kendengfauna. Tijdschr. Kon. Ned. Aardr. Gen. XXIV, p. 449—458.

1907 W. VOLZ. Das geologische Alter der Pithecanthropus-Schichten bei Trinil. Neues Jahrb. f. Min. Festband, p. 256—271.

1907 J. ELBERT. Ueber das Alter der Kendenggeschichten mit Pithecanthropus erectus. Neues Jahrb. f. Min. Beil. Bd. 25, p. 648—662.

1908 K. MARTIN. Das Alter der Schichten von Sondé auf Java. Kon. Akad. v. Wetensch. Amsterdam. Versl. Afd. Wis- en Natuurk. XVII, p. 7—16.

1908 E. DUBOIS. Das geologische Alter der Kendengsgeschichten oder Trinilfauna. Tijdschr. Kon. Nederl. Aardr. Genootschap XXV, p. 1235—1270.

1908 R. D. M. VERBEEK. Molukken-verslag. Jb. Mijnw. 1908, Wet. ged.

1908 J. ELBERT. De nieuwste onderzoekingen over het Pithecanthropusvraagstuk. Nat. Tijdschr. v. Ned. Ind. LXVII, p. 125—142.

1908 W. BRANCA. Vorl. Bericht über die Ergebnisse der Trinil-Expedition. Sitz. Ber. der K. Preuss. Akad. d. Wissensch. Phys. Math. Classe, p. 261—278.

1909 J. ELBERT. Dubois' Altersbestimmungen der Kendengschichten. Centralbl. J. Min, p. 513—520.

1910 R. A. DALY. Pleistocene Glaciations and the coralreefproblem. Amer. Journ. of Science XXX, p. 300.

1911 L. SELENKA. M. BLANCKENHORN. Die pithecanthropus-schichten auf Java. Geologische und Palaeontologische Ergebnisse der Trinil-Expedition (1907—1908). Including the following publications:

E. CARTHAUS. Zur Geologie von Java.

C. DOZY. Bemerkungen zur Stratigraphie der Sedimente in der Triniler gegend.

H. MARTIN-ICKE. Die fossilen Gastropoden.

K. MARTIN. Notizen über die Süsswasserbivalven aus den Pithecanthropus-Schichten von Trinil.

J. FELIX. Die fossilen Anthozoen aus der Umgegend von Trinil.

H. STREMME. Die Säugethiere mit Ausnahme der Proboscidier.

W. JANENSCH. Die Proboscidier-Schädel der Trinil-Expeditions-Sammlung.

E. CARTHAUS. Spuren von möglicherweise menschlicher Tätigkeit in den Trinil-Schichten.

J. Schuster. Die Flora der Trinil-Schichten.
M. Blanckenhorn. Algemeine Betrachtungen über die wissenschaftl. Ergebnisse der Selenka-Trinil-Expedition.
1911 H. Stremme. Die Säugethierfauna des Pithecanthropusschichten. Centralbl. f. Min., p. 54—60, 83—89.
1911 J. Schuster. Monographie der fossilen Flora der Pithecanthropusschichten.
1911 J. Elbert (and Hallier). Die Selenka'sche Trinil-Expedition und ihr Werk. Centralbl. f. Min., p. 736—741.
1914 K. Martin. Wann löste sich das Gebiet des Indischen Archipels von der Tethys? Samml. Keiden Ie Ser. Bd. IX, p. 337.
1914 W. Soergel. Stegodonte aus den Kendengschichten auf Java. Palaeontographica Suppl. IV (III, 1).
1915 R. A. Daly. The glacial-control Theory of coral reefs. Proc. of the Amer. Acad. of Arts and Sciences. LI, p. 173.
1916 L. M. R. Rutten. Vier dwarsprofielen over de tertiaire mergelformatie tusschen Soerabaja en Ngawi. Verh. Geol. Mijnb. Gen. Geol. Serie III.
1919 K. Martin. Unsere palaeozoologische Kentnisse von Java. Leiden.
1919 G. A. F. Molengraaff en Max Weber. Het verband tusschen den Plistoceenen ijstijd en het ontstaan der Soendazee (Java- en Zuid-Chineesche Zee) en de invloed daarvan op de verspreiding der koraalriffen en op de land- en zoetwaterfauna. Kon. Akad. v. Wetensch., Amsterdam. Versl. Afd. Wis- en Natuurk. XXVIII.
1920 R. A. Daly. A recent Worldwide sinking of Oceanlevel. Geologic. Magaz. Vol. LVII.
1920 P. Tesch. Jungtertiäre u. quartäre Mollusken von Timor. Palaeontologie v. Timor VIII.
1921 V. Hilber. Alter der Pithecanthropusschichten. Centralbl. f. Min. 1921, p. 149—154.
1221 P. Fischer. Eine Pliozänfauna von Seran. Centr. Bl. f. Min. 1921. p. 242—251, 278—286.
1923 R. Koch. Die Jungtertiäre Foraminiferenfauna von Kabu (Res. Surabaja, Java) Eclogae geologicae Helvitiae. Bd. XVIII, No. 2.
1923. I. van der Vlerk. Een nieuwe vindplaats van fossiele werveldieren op Java. De Mijningenieur. IV, p. 67—68.
1923 G. A. F. Molengraaff. De geologie der zeeën van Nederl. O. Indië, Leiden.
1924 E. Dubois. Over de voornaamste onderscheidende eigenschappen van den schedel en de hersenen, de onderkaak en het gebit van den Pithecanthropus erectus. Kon. Akad. v. Wetensch., Amsterdam. Versl. Afd. Wis- en Natuurk. XXXIII, p. 135.
1924 W. O. Dietrich. Zur Altersbestimmung der Pithecanthropusschichten. Sitz. Ber. der Ges. Naturforsch. Freunde.
1925 L. M. R. Rutten. Over de herkomst van het materiaal der neo-

gene gesteenten op Java. Kon. Akad. v. Wetensch. Amsterdam. Versl. Afd. Wis- en Natuurk. XXXIV, p. 689—708,
1925 L. M. R. Rutten. Over de richting der tertiaire gebergtevormende bewegingen op Java. Kon. Akad. v. Wet., Amsterdam Versl. Afd. Wis- en Natuurk. XXXIV, p. 65—78.
1925 I. van der Vlerk. Het Foraminiferengenus Spiroclypeus etc., Verh. Geol. Mijnb. Gen. Geol. Ser. VIII, p. 562—567.
1926 K. Martin. Pliocene versteeningen van Cheribon in Java. Dienst v. d. Mijnbouw in Ned. Ind. Wet. Med. No. 4.
1926 G. G. Stehlin. Fossiele Säugetiere aus der Gegend von Limbangan (Java). Dienst v. d. Mijnb. in Nederl. Ind. Wetensch. Med No. 3.
1926 J. Zwierzycki. De beteekenis van de nieuwe fossiele werveldierresten bij Boemiajoe. De Mijningenieur VII, p. 229.
1927 I. M. van der Vlerk en J. H. F. Umbgrove. Tertiaire gidsforaminiferen van Nederl. Oost-Indië. Dienst v. d. Mijnbouw in Ned. Ind. Wet. Med. No. 6.
1927 L. M. R. Rutten. Voordrachten over de geologie van Nederlandsch Oost-Indië. Groningen—den Haag.
1927 Davidson Black. The lower molar hominid tooth from the Chou Kou Tien deposit. Palaeontologica Sinica. Ser. D., Vol. VII, Fasc. 1.
1928 K. Martin. Eine Nachlese zu den neogenen Mollusken von Java. Leidsche Geol. Mededeelingen III, 2, p. 105—130.
1928 J. H. F. Umbgrove. De koraalriffen in de Baai van Batavia. Dienst v. d. Mijnb. in Ned. Ind., Wet. Med. No. 7.
1929 J. H. F. Umbgrove. Tertiary Sea connections between Europe and the Indo-Pacific Area. Proceedings Fourth Pacific Science Congress (Java).
1929 H. Gerth. The Stratigraphical distribution of the larger Foraminifera in the Tertiary of Java. Proceedings Fourth Pacific Science Congress (Java).
1929 C. ter Haar. Boemi-ajoe district. Excursion E4 of the Fourth Pacific Science Congress (Java).
1929 L. J. C. van Es. Trinil. Excursion E5 of the Fourth Pacific Science Congress (Java).
1929 Ch. E. Stehn and J. H. F. Umbgrove. Bijdrage tot de geologie der vlakte van Bandoeng. Tijdschr. Kon. Ned. Aardr. Gen., p. 301—314.
1929 Friedrich Siemon. Jung tertiäre Mollusken fauna aus Niederländisch Ost Indiën. Ber. d. Naturf. Gesellsch. zu Freiburg in Br. XXIX, p. 309.
1929 Davidson Black. Preliminary note on additional Sinanthropus material discovered in Chou Kou Tien, during 1928. Contributions from the Cenozoic Lab. of the Geol. Survey of China and the Department of Anatomy. Peking, Union Medical Coll. VIII. No. 1.

1929 Davidson Black. Preliminary notice of the discovery of an adult Sinanthropus skull at Chou Kou Tien. Bull. of the Geol. Society of China VIII, No. 3.

1929 W. C. Pei. An account of the discovery of an adult Sinanthropus skull in the Chou Kou Tien district. Bull. of the Geol. Soc. of China. VIII, No. 3.

1929 P. Teilhard de Chardin and C. C. Young. Preliminary report on the Chou Kou Tien fossiliferous deposit. Bull. Geol. Soc. of China. VIII, No. 3.

1930 Davidson Black. Interim report on the skull of Sinanthropus. Bull. Geol. Soc. of China. IX, No. 1.

1930 J. H. F. Umbgrove. De koraalriffen van den Spermonde Archipel. S. Selebes. Leidsche Geol. Med. III, 5, p. 227—247.

1930 C. W. A.P. 't Hoen. Geol. Overzichtskaart van den Nederl. Ind. Arch. 1 : 1.000.000. Toelichting bij blad XVI. Midden-Java. Jaarb. Mijnw. 1929. Verh. p. 1—62.

1930 ............. Javakarteering, Jaarverslag over 1929. Jaarb. Mijnw. 1929. Alg. p. 43—50.

1930 Tan Sin Hok. Over Cycloclypeus. Voorloopige resultaten eener biostratigrafische studie. De Mijningenieur.

1930 G. Elliot Smith. The ancestry of man. Bull. Geol. Soc. of China IX, No. 3.

1931 J. H. F. Umbgrove. De koraalriffen van Emmahaven (W.-Sumatra). Leidsche Geol. Med. IV, 1, p. 9—24.

1931 K. Martin. Wann löste sich das Gebiet des Indischen Archipels von der Tethys? (eine Fortsetzung). Leidsche Geol. Med. IV, 1, p. 1—8.

1931 J. Cosijn. Voorloopige mededeeling omtrent het voorkomen van fossiele beenderen in het heuvelterrein ten Noorden van Djetis en Perning. Verh. Geol. Mijnb. Gen. v. Nederl. en Kol. Geol. Ser. IX, p. 113—119.

1931 J. H. F. Umbgrove en J. Cosijn. Java's Zuidkust bij Tji-Laoet-Eureun. Verh. Geol. Mijnb. Gen. v. Nederl. en Kol. Geol. Ser. IX, p. 121—134.

1931 Davidson Black. On an adolescent skull of Sinanthropus pekingensis in comparison with an adult skull of the same species and with other hominid skulls, recent and fossil. Palaeontologia Sinica, Scr. D, Vol VII, Fasc. ii, Peiping.

Additional material from *The Age of Pithecanthropus,*
ISBN 978-94-017-5674-7 (978-94-017-5674-7_OSFO2),
is available at http://extras.springer.com

GPSR Compliance

*The European Union's (EU) General Product Safety Regulation (GPSR) is a set of rules that requires consumer products to be safe and our obligations to ensure this.*

*If you have any concerns about our products, you can contact us on ProductSafety@springernature.com*

In case Publisher is established outside the EU, the EU authorized representative is:

Springer Nature Customer Service Center GmbH
Europaplatz 3
69115 Heidelberg, Germany

**Batch number: 10824476**

Printed by Printforce, the Netherlands